CRITTERS

COMMON HOUSEHOLD AND GARDEN PESTS OF TEXAS

COMMON HOUSEHOLD AND GARDEN PESTS OF TEXAS

Bill Zak

TAYLOR PUBLISHING COMPANY

Dallas, Texas

Copyright © 1984, Taylor Publishing Company
1550 West Mockingbird Lane, Dallas, Texas 75235

Library of Congress Cataloging in Publication Data

Zak, Bill.
   Critters: common household and garden pests of
Texas.

   Includes index.
   1. Insects, Injurious and beneficial — Texas —
Identification. 2. Garden pests — Texas —
Identification. 3. Household pests — Texas —
Identification. 4. Insect control — Texas. 5. Garden
pests — Control — Texas. 6. Household pests —
Control — Texas. I. Title.
SB934.5.T4Z35 1984 628.9'657 84-8471
ISBN 0-87833-384-3

Designed by Bonnie Baumann

Printed in the United States of America

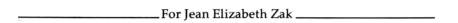

For Jean Elizabeth Zak

My wife, critic, helper, and avid supporter, who patiently shared her husband with a computer for several months while this book was written.

Acknowledgments

The manuscript portion of this book was reviewed for technical validity by Dr. Rodney L. Holloway, Extension Pesticide Impact Assessment Specialist, TAEX, Texas A&M University System. He also provided the information about scientific classification. I am forever grateful for his kindness, expertise, and encouragement.

I wish to offer most sincere thanks to the United States Department of Agriculture — Photography Division, Washington, D.C.; Van Waters & Rogers, division of Univar, San Mateo, California; Big State Pest Control, Houston, Texas; Texas Forest Service, a part of the Texas A&M University System; Southern Forest Insect Work Conference, Asheville, North Carolina; USDA Forest Service; Texas Agricultural Extension Service (Entomology Department) of the Texas A&M University System; Texas A&I University Citrus Center, Weslaco, Texas; Clemson University Extension Service in cooperation with the Federal Extension Service USDA, Washington, D.C.; and Ortho Information Services, Chevron Chemical Company, San Francisco, California. These public agencies and private companies made available the photographs shown in this book. I wish, also, to express appreciation for the efforts of innumerable entomologists and other employees of these agencies and corporations who caused to be published the myriad of technical documents that provided other data sources. And, my thanks to the Harris County Mosquito Control District, Houston, Texas, who provided mosquito data.

Individuals to whom I shall always be grateful: William J. Spitz, President, Big State Pest Control; Bruce R. Miles, Director, Texas Forest Service; Larry R. Barber, USDA Forest Service; Charles L. Cole, TAEX Entomologist; J. Victor French, Texas A&I University Citrus Center; John E. Fucik, Texas A&I University Citrus Center; John W. Norman, Jr., TAEX Extension Agent; Ronald F. Billings, Texas Forest Service; John N. Cooper, TAEX Extension Agent; Michael H. Shively, Harris County Extension Agent; Bette Lu Schwarz, President, ADKAR, Inc.; Sandi Cole, Student, Texas A&M University; Ray S. Reneria, Ortho — Consumer Products Division; Herbert A. Pase, Texas Forest Service; and Maggie (Grandma) and W. J. (Grandpa) Zak, my first teachers. My most sincere thanks to you all.

Bill Zak

Code for Slide Sources

C-USDA-xx	Clemson University Extension Service in cooperation with the Federal Extension Service USDA
ORTHO-xx	Ortho Information Services, Chevron Chemical Company
xxx-x-x VWR	Van Waters & Rogers, division of Univar
Ento.-TAEX	Texas Agricultural Extension Service — Entomology Department
IBB-x; SPB-x; HI-x; PD-x; TF-x; SI-x; RW-x; SC-x; WP-x	Southern Forest Insect Work Conference
TFS	Texas Forest Service
Big State-x	Big State Pest Control
John Doe	Private Individual

("x" indicates numbers used to identify individual slides within the series)

_____Table of Contents_____

_____ Introduction_____

My grandpa, Adolph Stasny, was a farmer, country store owner and cattle raiser near the small community of Smetana, Brazos County, Texas. A necessary family food source back in those days was the vegetable garden. Families were large and so were the gardens. My mother, known to some of you as Grandma Zak, tells how garden insects were controlled when she was a little girl some seventy-odd years ago.

When it got "good-and-dark," Grandpa gave each of the eleven children a kerosene lantern and a tin can with a little "coal-oil" in the bottom. Everybody took a row and, by lantern light, hand-picked every insect in sight, depositing same in their coal-oil can. It took quite a while to do this because the garden was somewhat over an acre in size. After the garden was scoured for insects, everyone turned in their cans to Grandpa who then dumped the critters in a pile and burned them. This ritual was repeated as needed. And those were the good 'ol days? I wonder how many of us would opt to garden if this technique were necessary today. Well, to coin a phrase, we've come a long way, baby!

The major problem in pest control is not the selection and use of a pesticide, but is rather the proper identification of the pest. Also keep in mind that some critters do noble deeds and should be spared. Which wear halos — which wear horns? This book is intended to help homeowners make those decisions. On the following pages, you will see suggested controls for the "bad guys." Be advised that these are intended to be only clues to controls. I don't suggest they are the only controls, or necessarily the best controls in any given situation. So, with that caution, let's proceed to meet our friends and foes. Welcome to the Critters, some common garden and household pests of Texas.

Bill Zak

Critters and Their Names

The world around us is filled with many different plants and animals. Creatures that vary almost beyond imagination in size, shape and color. Names are used to distinguish all these living things. Common names are used in our everyday conversation and scientific names are used for study. Most living organisms have two names, a common name and a scientific name. This book will use common names and also include enough scientific classification to give you an idea of how the animals relate to each other.

One of the most accepted classifications (naming systems) places all living things into one of two major groups — Plants and Animals. These are called Kingdoms and are further divided into Phylum, Class, Order, Family, Genus and Species. Using the American cockroach as an example, its classification would be as follows:

Kingdom — Animal
Phylum — Arthropoda
Class — Insecta
Order — Orthropoda
Family — Blattidae
Genus — Periplaneta
Species — americana

Almost all of the critters in this book are in the Animal Kingdom, the Phylum Arthropoda and Class Insecta. But there are a few exceptions: for example, rodents are in the Phylum Chordata, snails in the Phylum Mullusca, and earthworms in the Phylum Annelida. After the description of each critter is presented, those in the Class Insecta will have their Order and Family name listed. Some additional classification information will be presented for those outside the Class Insecta, but this will be kept within the scope of this book.

Dr. Rodney L. Holloway

___ Ants _____

CARPENTER ANTS

Carpenter Ants bore into moist, decaying wood, forming extensive galleries in which they make nests. They do not eat their sawdust-like wood borings, but feed on other insects, plant sap, pollen and seeds. Of over 2,500 known species of ants, carpenter ants and their relatives form one of the largest groups. When the colony grows too large, part of it will migrate, often invading nearby homes through windows, foundation openings, or similar entry points. They will bore into structural timbers, ceilings and floor or sub-floor areas, or they will colonize undisturbed hollow spaces such as wall voids and hollow doors where they will cause considerable damage if allowed to spread. In addition to weakening wood, carpenter ants may invade pantries, and often will inflict painful bites if disturbed. The workers, which are infertile females, are among the largest known ants; they grow up to ½ inch in length. Carpenter ants are black or reddish-black in color and may be winged or wingless.

Ants = Order Hymenoptera, Family Formicidae

_____ CONTROL CLUE _____

Spray baseboards, windowsills, door frames and other surfaces where these ants crawl with Diazinon or Baygon. Spray into the nests if possible. Remove nearby logs or stumps that might be an infestation source. Seal openings in the foundation, around windows and other access areas into your home. Don't be intimidated by the size of this ant; you can win! Go get 'em, tiger. Fight! Fight! Fight!

Carpenter Ant
(C-USDA-200)

Carpenter Ant damage
(Ortho-13)

Carpenter Ants
(Ortho-12)

FIRE ANTS

Fire Ants have become a tremendous problem for Texans. Four fire ant species are found in Texas: Tropical Fire Ants; Southern Fire Ants; Desert Fire Ants, although these are rarely observed; and Red Imported Fire Ants, which cause the most concern. Imported fire ants spread naturally through queen movement by crawling, by drifting downstream on or in logs, or by travelling aboard cars, trucks or trains. Shipments of soil or nursery stock from an infested area may relocate an entire ant colony or nest.

Agricultural losses from this pest are significant, but in urban areas imported fire ants are a formidable problem also. They invade lawns, parks, playgrounds, schoolyards, cemeteries, golf courses and homes. The ants attack anything that disturbs their mound. Symptoms of the fire ant sting include burning and itching, often followed by the occurrence of a white sore or pustule that may leave a permanent scar after healing. Venom of this ant is unlike that of any other stinging insect. Hypersensitive persons suffer chest pains or nausea and may even lapse into a coma from just one sting. Persons who react severely to a fire ant sting should see a physician immediately.

Ants = Order Hymenoptera, Family Formicidae

CONTROL CLUE

The technique used in applying an approved fire ant insecticide will be as important as the selection of the insecticide. To control any of the species effectively, an understanding of their habits, especially their mound-building activities, is critical. Fire ant mounds will be more apparent in the spring, summer and early fall following rainy periods. These ants do not prefer extremely wet soil; therefore, they increase their mound size and move their brood up into the drier soil of the mound. This habit makes a colony more vulnerable to insecticide control measures.

The most effective fire ant control is a bait application; workers hopefully will gather the poisoned food and feed it to the queen. Killing the queen is the key to colony elimination. Drenches do not always kill the queen, but if you drench properly, you will have increased your chances. The most effective drench procedure is to dilute an approved insecticide in water. Apply this solution (usually a gallon or so depending upon mound size) to simulate a gentle rain. A sprinkler can will be especially useful for this procedure. Thoroughly wet the mound and surrounding area to a diameter of 3 to 4 feet.

It is important not to disturb the mound before or during treatment. Mound disturbance causes part of the colony to move away from the chemical. If most of the colony is in the mound and above ground level, treatment should be successful. Surviving workers may remain for a week or two, but will ultimately die. If mounds are missed during the first application, re-treat when ants are noticed. This procedure is usually successful on cooler spring or fall days and during winter months. During very dry periods, most members of a fire ant colony are below ground level. Applying an insecticide at this time does little more than move the colony entrance to another location. To overcome the dry condition, water-soak the soil around the mound. This procedure drives the colony up into the drier soil. Now, proceed with control procedures previously described.

Several insecticides are labeled for fire ant spot treatment. They are available in liquids, baits and granules. Use according to label directions. I have had the most success with Orthene 75-W, which was recently cleared for use in Texas. Read the label carefully.

Imported Fire Ants gathering food
(IVA-10-C VWR)

Imported Fire Ant mound
(IVA-10-A VWR)

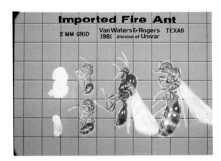

Imported Fire Ant larva, pupa, worker, drone & queen
(IVA-10-M VWR)

Typical Fire Ant mound in yard
(C-USDA-213)

Southern Fire Ant mound
(IVA-11-A VWR)

Southern Fire Ant in stinging position on an
insect
(IVA-11-I VWR)

HOUSEHOLD ANTS

Many different ants are commonly found in Texas and they readily invade homes for feeding or nesting or both. Homeowners become alarmed when worker ants are found infesting food or other items and ordinarily respond by drenching the area with enough insecticide to kill a mule. Applying extensive, undirected insecticide treatments in the home is unsatisfactory because this action kills only a small number of the invading workers and usually does little to affect the colony which is the source of the workers.

Ants come in your house because food, water and shelter are available there. They are one of the most successful insect groups because they so readily adapt to changing environments. Most ants prefer to nest in soil or outdoor wood, but your home offers many desirable locations for nesting; cracks and holes in brick veneer, wall voids and structural wood are sites commonly sought out and exploited. At certain times of the year, mature ant colonies produce winged reproductives which swarm from the nest in great numbers. When these swarmers emerge within or near a building, they may be confused with termite swarms, but if they are ants, the nest is usually located in proximity to the swarms.

Ants = Order Hymenoptera, Family Formicidae

The most effective ant control lies in finding the nest and treating it with the proper insecticide. S.O.S., SEEK OUT SOURCE! To locate ant nests, determine movement patterns and the worker ants will often lead you back to the nest. Another good technique is to use small bait stations to trick the ants into revealing their nest locations. Use soft drink or pill bottle caps baited with small amounts of peanut butter, jelly and bacon grease; one or more of these three foods will attract any ants. Watch the ants as they locate this food and begin taking it back to the nest. They may even establish an odor trail which other workers will use to come to the food source, so an ant column may develop. When you locate Ant City, USA, treat it with Sevin, Dursban, Diazinon or a Boric Acid Bait; the species of ant will determine which you use. Follow label directions.

Argentine Ant workers
(IVA-2-A VWR)

Little Black Ant worker
(IVA-13-A VWR)

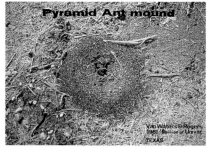

Pyramid Ant mound
(IVA-5-A VWR)

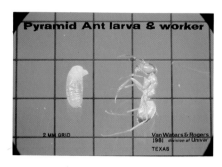

Pyramid Ant larva & worker
(IVA-5-H VWR)

Thief Ant worker
(IVA-12-A VWR)

PHARAOH ANTS

Pharaoh Ants are perhaps the most difficult household insect pests to control. They are quite small (less than 1/16 inch long) and are a light tawny-brown color. Pharaoh ants adapt well to living in homes because they can readily nest in wall voids, appliances, linens, heating ducts, light fixtures and attics. They prefer to be close to heat and moisture sources.

Pharaoh ants are commonly called "sugar ants," but I promise you their diet is more diversified. They will have a family reunion on a drop of bacon grease overlooked on the kitchen range. Draw your bath in the morning and find a hundred and fifty drinking water on the shower head. Brush your teeth and see that interminable trail coming from behind the mirror heading for lunch on your tooth paste tube. Put on the shirt you wore to dinner last evening and find them crawling pell-mell up the sleeves heading for that spot of salad dressing inadvertently splattered on your collar. Those cookies for the office left on the breakfast table will be the occasion for an orgy that would have awed the Romans. Welcome to pharaoh ants!

Ants = Order Hymenoptera, Family Formicidae

Pharaoh Ant workers tending eggs
(IVA-16-D VWR)

Pharoah Ant workers gathering food
(IVA-16-B VWR)

RED HARVESTER ANTS

Here's the ant that you will painfully, but nostalgically, remember if you grew up as a barefooted country kid — the "Red Ant" that would hunch-up on your big toe and inflict such a painful sting that it ruined your Sunday afternoon over at Grandma's house. This is the same critter that would make your dog yelp with pain if one happened to curl-up in a stinging knot between his toes. One red-ant sting would forever be a vivid reminder to respect that big, bare, sandy mound with the hole in the top. This is the home of the Red Harvester Ant ... where you dropped the dead snake, whose remains would mysteriously disappear after a couple of days.

Harvester Ants live in large, deep colonies with the queen ensconced 2 to 6 feet underground. These ants are also known as

Agricultural Ants because of their preference for seeds and other plant materials which foraging worker ants gather and haul into the nest with breakneck speed. All vegetation around their mound will be cleared, leaving big bare circles that may be 2 to 12 feet in diameter. Red Harvester Ants are ¼ inch or more in length.

Ants = Order Hymenoptera, Family Formicidae

CONTROL CLUE

These are "country" ants and they seldom require control, but if their presence poses a danger to animals or humans, or if their mound proves to be undesirable, treat the colony with Methyl Bromide, as you would for the Texas Leaf Cutting Ant. Read the label.

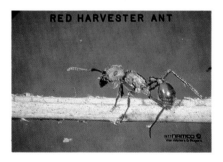

Red Harvester Ant
(IVA-7-A VWR)

Red Harvester Ant mound
(IVA-7-I VWR)

Black Harvester Ant
(IVA-8-A VWR)

TEXAS LEAFCUTTING ANTS

Texas Leafcutting Ants are also commonly known as town ants, cut ants, parasol ants, fungus ants or night ants. They occur principally in East Texas, living in large colonies with populations that may exceed 2 million ants. Leafcutting ants share the sophisticated habit of growing a "fungus garden," as do certain termites in Asia and Africa and a few of the wood-boring beetles. Leafcutting ants remove leaves and buds from weeds, grasses, peach and plum trees, blackberry bushes, vegetable gardens and many other fruit, nut and ornamental plants as well as several cereal and forage crops. In East Texas and West-central Louisiana, young pine seedlings are often destroyed within a few days if planted where these ants are abundant.

These big, brown, worker ants are active from May to September, foraging during the night. Common worker ants range from ¼ to ½ inch in length. They often completely defoliate plants, carrying severed leaves to their nest to maintain their fungus garden which eventually is used for food. They travel along well defined, multi-lane trails, rushing pell-mell to unload a hunk of leaf; then back they come, bumper-to-bumper, at the same pell-mell pace, to get another load. Watching these "truck drivers" of the insect community will be completely fascinating, until you realize that they are hauling off your garden.

Ants = Order Hymenoptera, Family Formicidae

CONTROL CLUE

Follow that truck driver! The trail will lead you to the colony which may consist of only a few small mounds or one mound which may be several feet across. Methyl Bromide will be the best control, and be sure you find out how to use it if you don't know. Methyl Bromide is tough stuff.

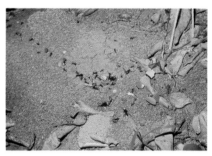

Texas Leafcutting Ant trail
(Dr. J. Victor French)

Texas Leafcutting Ant on leaf
(Dr. J. Victor French)

Texas Leafcutting Ant colony
(HI-40 SFIWC)

_____ Antlions (Doodlebugs) _____

When you were a child, did you ever play with a Doodlebug? Every Texas kid must know this critter in order to experience a normal, healthy childhood. If you don't know about Doodlebugs, you must be from "out-of-town." Let's proceed with your education — it's never too late to do the right thing.

A Doodlebug is an Antlion and is so named because the larvae of this family have strange eating habits. Adults resemble miniature damselflies. Females lay eggs on the ground in sand or sandy soil. When one hatches, the larva digs a pit shaped like an inverted cone with unbelievably smooth interior sides, and this larva lives almost completely buried at the bottom (the tip of the cone). This abode is a superbly designed trap, and should an ant or other small insect tumble in, it is seized in the powerful jaws of the larva and is sucked dry. Touch the upper edge of this sandy-sided trap with a twig and watch the action in the bottom of the hole. This guy does the original "jitter-bug," causing a sand-slide which will carry the prey into those waiting jaws. As the larva matures, it builds a silken cocoon in which it pupates. Doodlebugs are harmless (except to the critter that falls into the hole).

Note to Mothers of Small Children: If you want a free baby-sitter for an hour, take li'l darlin' out in the back yard and find a doodlebug house. Of course, baby will drive that doodlebug crazy, but that's all part of being a doodlebug.

Antlions = Order Neuroptera, Family Myrmeleontidae

BENEFICIAL

Antlion (Doodle Bug)
(C-USDA-184)

_____ Aphids _____

When your young and tender garden plants begin to look puny for no apparent reason, you'd better begin checking for aphids. Aphids are small, sluggish, soft-bodied insects often called plant lice. Sometimes they are referred to as Ant Cows, and I'll tell you why in just a minute.

Aphids come in various colors, like black to green to yellow and various shades in between. These fellows are suckers in the true sense of the word. A number of species attack various field and garden crops. You will find them on trees . . . on bushes . . . on potted plants, congregating on the new tender growth, sucking plant sap, causing stunting and leaf curling and leaving that most undesirable honeydew deposit.

Certain ants caress aphids, inducing them to produce more honeydew, which the ants eat. If aphids become crowded on a plant, the ants will often move some aphids to uninfested plants in order to supply them with a more ample food supply. They will tend them like a herd of cattle . . . hence, the name Ant Cows. Incidentally, aphids give birth to living young and breed prolifically. Some are winged and will fly.

Aphids = Order Homoptera, Family Aphididae

Aphids tended by ants
(C-USDA-106)

Aphids (Plant Lice)
(C-USDA-95)

Aphids and predator
(C-USDA-66)

Spider predator feeding on aphid
(100-1-FF VWR)

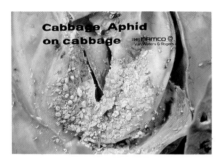

Cabbage Aphids on cabbage
(HA-7-A VWR)

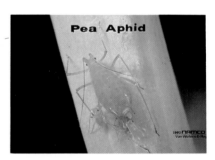

Pea Aphid stem mother giving birth
(HA-25-A VWR)

CONTROL CLUE

Some gardeners simply wash them off a plant with the garden hose, but better you should use a shot of Malathion or Diazinon. Fortunately, aphids aren't difficult to control.

Armyworms

To be invaded by a horde of Armyworms is to know how Custer felt at Little Big Horn. Armyworms get their name from their feeding habits. After they have eaten everything in one area, they crawl in droves to another area in search of more food. Armyworms are caterpillars. After several weeks of feeding, they pupate in the soil, then emerge as adult moths to repeat the cycle.

Several varieties of armyworms attack the home garden. Among them are the Fall Armyworm, Yellow-striped Armyworm and Beet Armyworm. In South Texas, all stages of the beet armyworm may be found throughout the year. Fall armyworm moths can overwinter along the Texas Gulf Coast and then in the spring fly north to do their dirty-work. Larvae of the yellow-striped armyworm generally are day feeders on foliage of forage plants, but otherwise their habits are similar to their armyworm cousins. They'll go after anything green and try to eat all of it. Armyworms may migrate from lawns, pastures or small grains into vegetable gardens.

One year in Jackson County, I saw armyworms so bad you could actually hear 'em. And if I didn't hear 'em, I thought I did. Sounded like cottonwood leaves rustling in a soft breeze.

Armyworms = Order Lepidoptera, Family Noctuidae

Fall Armyworm having lunch
(C-USDA-85)

Yellow Striped Armyworm who already had a bite
(C-USDA-50)

Beet Armyworms about to have lunch
(Ento.-TAEX)

———————— CONTROL CLUE ————————————

In your garden use Sevin or Diazinon and hurry!

———— Assassin Bugs ————————

Assassin Bugs come in several different models. They are sometimes brightly colored, sometimes less so. They present an ominous manner, but are actually beneficial. Assassin bugs are very similar in appearance to leaffooted bugs and the adults are winged.

If ever you are invited to lunch by one of these friendlies . . . don't go. A favorite meal consists of chomping into a big ol' caterpillar and then proceeding to suck the body juices from it. They probably do this same number on other insects also. WOW! What a way to live, but to each his own, I guess.

Assassin Bugs = Order Hemiptera, Family Reduviidae

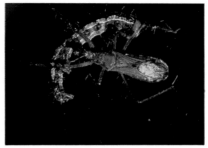

Assassin Bug adult
(HE-14-E VWR)

Assassin Bug having caterpillar pie
(Ento.-TAEX)

_____ Bagworms _____

Bagworms are a common pest of ornamental trees. They are most commonly found on arborvitae and juniper, but they also will prey on maple, box elder, sycamore, willow, black locust, elm, basswood, poplar, oak, apple, cypress, wild cherry, sassafras, persimmon and cedar. Wow, what a menu! Bagworms have the unusual habit of spending their entire larval stage inside a silken bag. It is very strong and is well camouflaged with bits of twigs and leaves from the host plant. The early stages of an infestation are difficult to detect. As the larvae grow, the most obvious evidence of their presence is the defoliation of the upper portion of the tree and the presence of those Christmas-tree-like ornaments attached to the twigs and leaves where larvae are feeding.

Another unusual trait of the bagworm is that the adult female never leaves her bag. She is grub-like in appearance, has no wings or eyes and has no functional legs, antennae or mouth parts. Her body is soft, creamy-white and is nearly hairless. (No wonder she stays in the bag; how would you like to be married to that?) Large numbers of arborvitae and juniper are killed each year as a result of complete defoliation by bagworms.

Bagworms = Order Lepidoptera, Family Psychidae

CONTROL CLUE

Natural controls play an important role in keeping populations at tolerable levels. Bagworms may be killed by a variety of wasp parasites, by birds that prey on the young larvae and by low winter temperatures. Sprays should be applied when young larvae first appear in the spring. Isotox is a good spray. Read the label.

On small yard and ornamental plantings, a simple method of control is to hand pick the bags and burn them immediately. Don't leave a bucket of bagworms out in the garage to be disposed of the next day. Those guys will crawl out of that bucket overnight, and you will be re-picking bagworms off of your car, the walls, the ceiling and everything else inside your garage. Listen to this voice of experience.

Bagworm close up
(C-USDA-126)

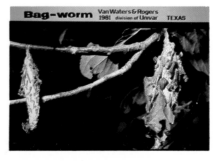

Bagworm cocoons on plum tree
(LM-16-A VWR)

Bark Lice

The Bark Louse prompts more questions than the Tooth Fairy. Every year, especially during early fall, my partner, Ben Oldag, and I receive predictable calls and letters from worried KTRH listeners asking about a mysterious, silvery web that covers the trunk and limbs of a favorite tree. This phenomenon occurs suddenly, frequently overnight; you walk outside one morning to get the paper and there it is, that eerie white web, coating the tree. Surely this is the work of an evil force from outer space, or an insidious plague that will surely destroy the tree!

Don't worry. This is the work of the bark louse who, with his family, has come to do you a favor. These tiny creatures are scouring the bark for plant, animal or fungus life which they eat, and while they perform this house-keeping chore, they spin this web for protection. Usually the web will disintegrate in two or three days and will disappear almost as suddenly as it appeared. *No control is necessary.* You could blow the web away with water pressure from your garden hose, but don't you dare. These little guys have it tough enough in trying to clean up that tree trunk for you.

Bark Lice = Order Psocoptera, Family Pseudocaeciliidae

BENEFICIAL

Bark Louse on tree bark
(PS-1-A VWR)

Bark Louse web on tree in Houston
(Mike Shively)

Bees

HONEY BEES

Honey bee is everybody's friend. Although he is wrongfully blamed for many stings that are actually performed by the aggressive "yellow jacket" paper wasp, this popular insect is the prime practitioner in the insect world of the old philosophy . . . Live and Let Live.

Honey bee makes honey. You know that. But do you know that the major contribution the honey bee makes to mankind is the function of plant pollinator? This industrious critter is definitely beneficial, but he will pop it (stinger) to you if you mess with him. Don't fool around his house (hive), and woe be unto your bottom if you sit on him. Every summer I lift a number of honey bees out of our swimming pool with my bare hand. But don't try this with your thumb and forefinger. This treatment is a squeeze and he'll sting you. Lift him out with the palm of your hand, and tump him out on the grass. Honey bee will buzz his wings several times to dry them and off he will fly to do his chores, and you will have done your good deed for the day. If you're out in the garden and one buzzes around your face, don't swat at him. He doesn't deserve that. Stand perfectly still. He will satisfy his curiosity and fly away. If perchance he sits on your nose, for god's sake don't slap him . . . and you know why.

So, now you have a few tips on how to get along with Honey bee. Remember, be nice. Live and let live. And eat more honey!

Honey Bees = Order Hymenoptera, Family Apidae

BENEFICIAL

HONEY BEE POLLINATION

Honey Bee adult
(III-11-A VWR)

WILD BEES

Honey bees have some kinfolks. There are several species, but let's save time by referring to them simply as the Wild Bees. Adult colors will vary, but blacks, oranges and yellows are the most common. Some are metallic green or blue.

Wild bees nest in a variety of locations — and by the way, not all bees live in colonies; many live alone. BUMBLE BEES nest in the ground and their lifestyle is simple; they won't bother you if you don't bother them. CARPENTER BEES are common in Texas and look very much like bumble bees, but they nest differently. Carpenter bee activity is commonly found around wooden fences, patio covers, wood shingles, roof eaves, porch ceilings, doors and windowsills. In establishing a nesting site, the female bores into wood at a right angle to the surface. The characteristic entrance tunnel is clean cut, ½ to 1 inch deep, and is approximately ½ inch wide — about the size of a dime. In many areas of Texas there may be two or more generations a year, but in South Texas continuous generations may occur. Damage to structures is usually slight and amounts to cosmetic defacement, but try to sell that line to your wife when she finds that 45-caliber "bullet-hole" out by the back door.

The phantom-like LEAF CUTTING BEE is blamed for stealing those leaf hunks off your rose bush. My friend and prominent Rosarian, Baxter Williams, insists (in jest) that the leaf cutting bee is merely the figment of entomologists' imagination and doesn't really exist. Baxter contends he has never seen a leaf cutting bee at work. Not many people have. This bee is indeed a solitary critter who nests in the ground and uses those leaf hunks to line its nest.

Faults notwithstanding, wild bees also do an excellent job of pollinating. Some native bees are better pollinators than honey bees. Credit where credit is due, folks. Let's have a round of applause for the Wild Bees.

Bumble Bees = Order Hymenoptera, Family Apidae
Carpenter Bees = Order Hymenoptera, Family Xylocopidae

CONTROL CLUE

Remember, these Wild Bees are beneficial, but if you are pushed to a control for the carpenter bee, use liquid Sevin applied into the nest entrance and on a wide area of wood surface around the entrance hole. Wait 12 to 24 hours after application, then plug the hole. New adults will continue to emerge if insecticides are not used before plugging. Take care. Females are very aggressive, particularly during nesting activity. Bumble bees are treated more often than carpenter bees. Treat a bumble bee's nest by pouring the insecticide down the entrance hole, but remember they have a front door and a back door, so be sure you plug one before you treat the other or you're gonna get a painful surprise.

Bumble Bee adult
(III-12-A VWR)

Valley Carpenter Bee/Bumble Bee comparison
(IX-13-G)

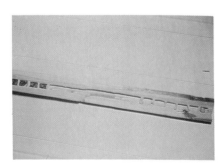

House damage from Carpenter Bee
(WP-22)

Leaf Cutting Bee cutting leaf
(Ento.-TAEX)

Leaf Cutting Bee damage to rose
(HY-18-A VWR)

Leaf Cutting Bee close-up
(Ento.-TAEX)

_____ Beetles _____

BLISTER BEETLES

Several species of Blister Beetles do business in Texas. The adults are long and slender with distinct body divisions and may be black, gray or striped. These fellows aren't all bad — just mostly bad. The larvae of some species are beneficial because they feed on grasshopper eggs. The adult beetles, however, eat plant leaves and can defoliate entire plants. They especially enjoy potatoes, tomatoes, eggplants, beans and peas, along with several other vegetables.

Be careful when handling these bugs; the juice from crushed beetles causes blisters. Guess how the Blister Beetle got its name!

Blister Beetles = Order Coleoptera, Family Meloidae

_____ CONTROL CLUE _____

Like the other beetles — Sevin or Diazinon. But remember . . . don't touch.

Striped Blister Beetle
(C-USDA-40)

Margined Blister Beetle
(C-USDA-41)

Blister Beetle (black)
(CO-10-A VWR)

COLORADO POTATO BEETLE

Colorado Potato Beetle adults are husky, yellow and black striped beetles that are about ⅜ inch long. These critters are also known as Potato Bugs, but will often devastate tomato, eggplant and pepper plants as well. About 500 eggs are deposited in batches of 24 or so on the underside of leaves. The eggs hatch in 4 to 9 days and the larvae become full grown in 2 to 3 weeks. Larvae are reddish in color and are humped; they also are equipped with two rows of black dots on each side of the body. Two to three generations per season occur in Texas.

This pest was once a bane of commercial potato growers and still causes considerable damage in home gardens. Both adults and larvae feed on leaves and can completely strip a plant in short order.

Colorado Potato Beetles = Order Coleoptera, Family Chrysomelidae

---

CONTROL CLUE

Use a dust formulation of Sevin or Diazinon when these insects and damage are first noticed. Repeat every week till the infestation is eliminated.

Colorado Potato Beetle adult
(C-USDA-94)

Colorado Potato Beetle larva
(C-USDA-93)

CUCUMBER BEETLES

Two Cucumber Beetle varieties deserve our attention: Spotted Cucumber Beetles and Striped Cucumber Beetles. Mrs. Spotted, infamous in her own right, must also bear the blame for her progeny, Southern Corn Rootworms. (Reminds me of Ma Barker and her desperado sons.)

Both Striped and Spotted cucumber beetle adults are guilty of the same sins when it comes to goofing up your garden . . . like, they eat plants. Some favorites are cucumbers, squash, pumpkins and melons. They will even feed on stems and young leaves of seedlings soon after they emerge. Also while gadding about your garden, they spread two serious diseases of cucurbits — Mosaic and Bacterial Wilt. Their grubby kids (larvae) eat stems and roots below ground line, causing attacked plants to wilt or to be stunted. We'll discuss rootworms in more detail later.

Cucumber Beetles = Order Coleoptera, Family Chrysomelidae

```
 _____ CONTROL CLUE _____

  Use Sevin or Diazinon. Treat at first sign of the beetles.
  Repeat at weekly intervals as plants become reinfested. Don't
  tarry! When you control spotted cucumber beetles, you are
  also controlling southern corn rootworms.
```

Twelve-spotted Cucumber Beetle
(CC-5-A VWR)

Striped Cucumber Beetle and feeding damage
(C-USDA-90)

ELM LEAF BEETLE

The Elm Leaf Beetle can be found in Texas wherever elms grow, and 2 to 4 generations of this pest may occur per year. Adult beetles are about ¼ inch long; larvae (worms) are about ½ inch in length. Adult beetles eat holes in the leaves, but the larvae cause most of the damage by skeletonizing the leaf surface. If large populations develop in residential areas, the beetles may become a further nuisance as winter approaches, because they will invade homes.

These beetles will begin to leave their protected overwintering sites in the spring in order to fly to nearby elm trees at about the time elm leaf buds begin to swell. Adults feed by chewing holes in the unfolding leaves. Egg laying begins a short time later. Masses of up to 25 yellow-orange eggs are deposited on the underside of leaves and these eggs will hatch in about a week. The larvae will then feed on the under-surface of the leaves, leaving only the veins and the upper leaf surface. Those damaged leaves soon dry, turn brown, and may drop from the tree. When these conditions are severe and trees are without leaves for several consecutive years, limbs or perhaps the entire tree may die.

Elm Leaf Beetle = Order Coleoptera, Family Chrysomelidae

_____ CONTROL CLUE _____

Di-Syston 15% granular systemic insecticide may be used if you apply it early enough to be assimilated by the tree, or a Sevin 50-W spray will be quicker if the rascals slip up on you. Also note: if the tree is old and large, a systemic treatment might not be practical.

Elm Leaf Beetle adult
(C-USDA-146)

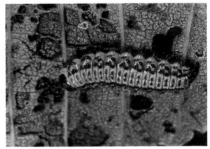

Elm Leaf Beetle larva
(C-USDA-147)

FLEA BEETLES

Flea Beetles are some of the most annoying of all vegetable pests. As beetles go, these guys are runts. Adults are approximately 1/16 inch long and are equipped with enlarged flea-like hind legs. As a matter of fact, flea beetles jump like fleas, but are not related to them. Some are striped, but most are either black, brown or green. Larvae are small, slender and white with a black band, and they have three pairs of legs.

Flea beetles attack numerous plants, but preferred vegetable hosts are tomatoes, peas, cabbage, carrots, eggplants, turnips, mustard, potatoes, beans and others. Both adults and larvae cause plant damage. Adults chew numerous, tiny, rounded or irregular holes in foliage so that leaves appear to have been peppered with fine shot. When feeding damage is heavy and there are many holes, leaves may wilt and turn brown; the host plant may become stunted and may even die. The larvae feed on roots or tubers. Life cycle from egg to adult may be completed in 6 weeks or less. One to four generations per year may develop.

Flea Beetles = Order Coleoptera, Family Chrysomelidae

CONTROL CLUE

Don't let these midgets fake you out. They are easily overlooked because they are so small. Watch for those shot-holes in the leaves and then get after them quickly with a spray containing Sevin, Malathion or Diazinon.

Flea Beetle adult on radish leaf
(CC-2-B VWR)

Flea Beetle damage
(Ento.-TAEX)

GOLDEN TORTOISE BEETLE

Golden Tortoise Beetle adults are oval, flattened, golden in color and are nearly ¼ inch long. Larvae are short, flattened and margined with a forked posterior appendage bent foreward over the body. This appendage holds a mass of cast skins and excreta. (Ugh!)

Both adults and larvae feed on the foliage of eggplant, sweet potato and other plants in the morning-glory and night shade family. These critters cut holes in leaves and sometimes will consume the entire leaf. There are other varieties of this beetle with different markings.

Tortoise Beetles = Order Coleoptera, Family Chrysomelidae

CONTROL CLUE

Sevin or Diazinon are usually adequate remedies for this "gold bug." Remember, read the label.

Tortoise Beetle (Gold Bug), two species
(C-USDA 103)

GROUND BEETLES

Several species of Ground Beetles prey on a variety of other insects, insect eggs and other anthropods (critters). One species, the Seed Corn

Beetle, will occasionally damage germinating seeds or seedlings, but this is an exception. Considering their many other attributes, we might readily score them 98% beneficial.

Adults are normally black. Some species may be brown, or may have red legs, or may have iridescent green-blue elytra (wing covers). They are very active and are fast movers. Larvae are fleshy grubs with large jaws. They generally remain in the soil, so you're not likely to see them. If you accidentally found one, I think you would kill it, because it is such an ugly thing. Shame, shame on you.

Ground Beetles = Order Coleoptera, Family Carabidae

_____ **BENEFICIAL**_____

Ground Beetle (Fiery hunter)
(C-USDA-186)

Ground Beetle (Caterpillar hunter)
(CO-1-A VWR)

Tiger Beetle
(C-USDA-193)

LADYBIRD BEETLES

Here's the darling of the insect world, commonly called "Lady Bug." A real sweetheart! Typically colored red with black spots, she may also be black or black with red spots or even gray with black spots.

Ladybird and her children (larvae) use chewing mouthparts to feed on aphids and other small insects that plague your garden. Learn her markings well lest you confuse her with the Mexican Bean Beetle or the Squash Beetle or one of the other bad guys and in error wipe her out. Mother Nature cursed Ladybird in another way . . . gave her UGLY children. You will wonder, can anything that ugly be good? Oh, well, you can't have everything.

Ladybird Beetles = Order Coleoptera, Family Coccinellidae

_____ **BENEFICIAL**_____

Ashy Gray Ladybird Beetle adult
(HP-1-L VWR)

Orange Ladybird Beetle adult
(C-USDA-190)

Black Ladybird Beetle on pine scale
(HS-24-J VWR)

Ladybird Beetle larva after aphids
(C-USDA-189)

MEXICAN BEAN BEETLE

I guess the Mexican Bean Beetle is how Mexico got even with Texas for the Battle of San Jacinto. The Mexican bean beetle looks very much like the ladybird beetle. They are relatives, but in function the similarity ceases.

The Mexican bean beetle has sixteen black spots on its back. Its basic color is usually copper-like or orange. Size is like that of the ladybird, about ¼ inch long. Orange to yellow soft-bodied grubs about ⅓ inch long with black-tipped spines on their back may also be present. These are the larvae (babies). These critters primarily work on the underside of your bean leaves. The leaves appear to dry up and the plant may die.

You're likely to tolerate the Mexican bean beetle thinking it to be a ladybird, or you will wipe out the ladybird, thinking it's the Mexican bean beetle. Pay attention.

Mexican Bean Beetle = Order Coleoptera, Family Coccinellidae

CONTROL CLUE

Diazinon or Sevin. Be sure to apply to the underside of the leaf. And be sure you have the right critter; the ladybird beetle is a beneficial insect and should not be eliminated.

Mexican Bean Beetle family
(C-USDA-96)

PINE BARK BEETLES

Five species of Bark Beetles are responsible for most pine damage. They are the SOUTHERN PINE BEETLE, the BLACK TURPENTINE BEETLE and the three southern IPS ENGRAVER BEETLES. More pines are killed by bark beetles than by any other group of insects. The southern pine beetle is by far the most destructive. Outbreaks may cover many acres and kill thousands of trees, or simply wipe out that favorite pine in your front yard.

A tree may be killed by the attacks of a single species of bark beetle, but it is very common for two or more species to attack the same tree. Ips beetles may be attracted to trees attacked by the southern pine beetles and vice versa. Trees heavily attacked by these beetles are doomed to die. Bark-beetle injured trees can be difficult to detect at an early stage because the small, yellowish-white or reddish-white masses of resin which mark the beetles' entry are often inconspicious. The foliage of heavily infested trees changes from dark green to light green, then to yellow, then to sorrel and finally to reddish brown. If trees are infested in late fall, they may not fade until the following spring.

Bark Beetles = Order Coleoptera, Family Scolytidae

CONTROL CLUE

Prevention is the only cure. Sorry! Maintain healthy trees. *Don't stress them* by building sidewalks or patios over the roots or by digging excessively around the roots, etc. They may be protected for 2 to 6 months by spraying with Lindane or Dursban-4E. Add a surfactant for better penetration of the insecticide, and spray until the solution runs down the bark crevices. Infected trees should be sprayed or cut and burned in order to protect healthy trees.

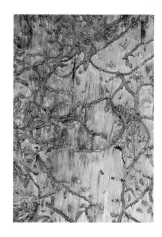

Adults of the 5 common species of Bark Beetles that attack and kill pines. The largest is the Black Turpentine Beetle; next the Southern Pine Beetle; in the middle is the Coarse Writing Pine Engraver; then the Southern Pine Engraver; then the small Southern Pine Engraver.
(IBB-13)

Symptom of SPB attack: "S"-shaped egg galleries produced by female SPB beneath bark of attacked tree
(SPB-14)

Southern Pine Beetle adult in gallery
(SPB-17)

Adult SPB covered with mites
(SPB-28)

Clerid Beetles, predator of SPB
(SPB-24)

Temnochila Virescence, predator of SPB
(SPB-26)

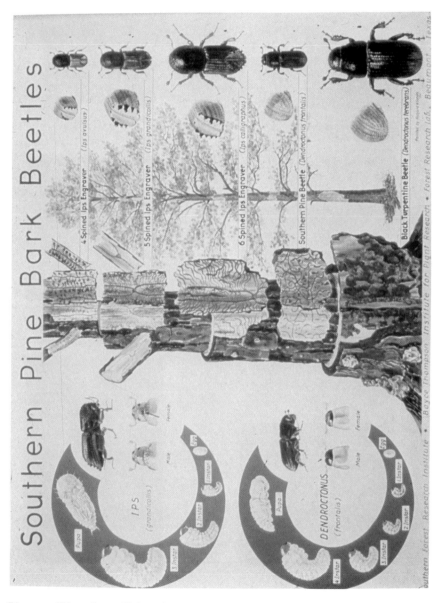

Diagram of life cycles and infestation patterns of 5 species of Southern Pine Bark Beetles
(IBB-3)

IPS (Ambrosia) Beetle boring
dust at base of tree infested by
SPB
(SPB-19)

Bark beetle pitch tubes, including Southern
Pine Beetle and Black Turpentine Beetle
pitch tubes. Also Pine Engraver entrance
hole with and without pitch tubes.
(IBB-23)

Ambrosia Beetle larva, pupa and adult in
galleries
(IX-12-C)

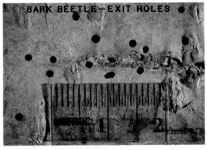

Ambrosia Beetle exit holes
(IX-12-E)

Larva of the predator Clerid working
through the galleries bored by the SPB,
searching for prey
(IBB-16)

Symptom of SPB attack:
boring dust in spider webs at
base of pine
(SPB-11)

PINE COLASPIS BEETLE

Colaspis Beetles seem to prefer slash pine as a host, but they have been found on many other southern pine trees. Feeding damage caused by large populations occasionally produces a spectacular browning effect of the needles similar to that caused by fire damage. Adult colaspis beetles chew the edges of needles to produce an irregular, saw-like edge which turns brown. Initially only the tips of the needles show these signs of infestation, but later the entire needle may die, causing the whole tree to turn brown and appear to be dying. However, attacked trees do not die, and little or no growth loss results. This pest is sporadic in its occurrence and may not develop again in the same area for several years. Attacks usually occur in early summer, but by late summer the trees appear green and healthy again.

The adult female colaspis beetle lays her eggs in the soil during the summer. Larvae hatch and feed on roots of grasses and other vegetation, and overwinter in this stage. The larvae pupate in the spring; adults emerge in early summer to feed. There is only one generation a year.

Pine Colaspis Beetle = Order Coleoptera, Family Chrysomelidae

---

CONTROL CLUE

Under forest conditions, no control measures are recommended for the pine colaspis beetle, but in your yard, practical control may be attained by using a Lindane spray.

Pine Colaspis Beetle and damage to loblolly pine
(PD-12)

SOFT-WINGED FLOWER BEETLES

Remember the red cross! And don't say "I gave at the office." The red cross that I refer to might be a characteristic marking on the back of a Soft-winged Flower Beetle. Several related species are found in gardens. This friend is about ¼ inch long and can usually be found on a bright, sunshiny day scurrying about on plants or flowers. This "Speedy Gonzales" preys on almost any smaller insect. This critter's fast footwork enables him to cover so much plant territory that he probably does a better job of pollinating than we give him credit for.

Soft-winged Flower Beetles = Order Coleoptera, Family Melyridae

 BENEFICIAL

A Soft-winged Flower Beetle
(CO-5-A VWR)

TWIG GIRDLER

A Twig Girdler is a beetle that does dirty-work on your trees by pruning off limbs. They could well be arrested for Malicious Mischief, but there is motive to this wood-carving madness.

Female twig girdlers do this unauthorized pruning after laying eggs in the limb part that is ultimately severed. Girdling begins during late summer, but the damage may not be noticed until the fall when the wind causes the withered branch to snap off and fall to the ground. It's all part of the twig girdler master plan for procreation. The eggs hatch into larvae which grow in the fallen limb. The larvae ultimately emerge, resulting in more mamas which will in time cut off more limbs.

Twig girdlers will attack apple, ash, dogwood, elm, hickory, mimosa, oak, peach, pear, pecan and other shade trees.

Twig Girdler = Order Coleoptera, Family Cerambycidae

CONTROL CLUE

If you want to try chemicals, use a Lindane spray when girdling is first noticed. Three applications at two-week intervals are usually necessary to prevent further damage, but that's a lot of trouble for questionable results. So, try this remedy. If you can catch mama girdler in the act and if you can reach her, bash her with a club. Next, gather any cut limbs that may be on the ground and burn them; there will be fewer twig girdlers next year. True, the damage will have been done, but just think, all that gnawing for nothing. Fight dirty!

Twig Girdler and damage to limb
(C-USDA-168)

WOOD DESTROYING BEETLES

POWDERPOST BEETLES attack only seasoned hardwood and generally feed on seasoned wood. Thus, powderpost beetles found in hardwood floors or furniture would not be expected to also attack the softwood (e.g., pine) structural timbers in a home. (On the other hand, DEATHWATCH BEETLES can attack both softwoods and hardwoods and also generally feed on seasoned woods.) One of the most significant wood infestors is the OLD HOUSE BORER, which generally attacks structural softwoods. Contrary to its name, it is often found in newer homes built with infested wood and will readily reinfest structural timbers. Wood that is improperly kiln-dried or treated, or wood that is stored too long is more likely to be attacked.

There are several indications that wood-boring beetles are present in a wood member. Immature beetles tunneling in wood cause an audible rasping or ticking sound most often heard during quiet times at night. Another indication may be a blistered appearance on the wood caused by larvae (worms) tunneling just below the wood surface. In feeding, beetles often push powdery frass (sawdust-like stuff) through holes made in infested wood. The consistency of the frass ranges from fine to coarse, depending on the species. Entry or exit holes without the frass may also be observed. Occasionally, wood staining or the actual sighting of adult beetles will be noted. Adult beetles emerging from their havens are often attracted to lights or windows.

The majority of flatheaded borers, roundheaded borers, bark beetles and timber worms are found shortly after a structure is built. Adults of these species generally will emerge within a few years after a building has been constructed and will not reinfest other wood. There are exceptions, of course, such as the old house borer. Proper identification is extremely important since controls vary according to species.

Powder Post Beetles = Order Coleoptera, Family Lyctidae
Death Watch Beetles = Order Coleoptera, Family Anobiidae
Old House Borer = Order Coleoptera, Family Cerambycidae

CONTROL CLUE

Prevention is the best control. Carefully inspect antique furniture, picture frames, bamboo products and other wood items before buying them. Consider any evidence of emergence holes, larval infestation or frass presence to be an indication of active infestation. Store fireplace wood as far from home as possible and only bring in firewood that will be immediately burned. Adult beetles can emerge from wood stored in the home and infest structural wood or furniture.

Small home furnishings, wooden artifacts, or small furniture pieces may be treated by freezing, but it is necessary to maintain the items at 0 degrees F. for several weeks to effect control. A localized infestation may be treated by applying a residual insecticide such as water-emulsified Lindane. Two coats are desirable, with the second being applied before the first completely dries. Proper treatment of unfinished wood may be effective for 10 years or more, but this will eliminate only adults that emerge or that attempt to reinfest. Larval development beneath the wood surface often continues and deep boring beetles such as the old house borer are usually not reached.

Fumigation is the most reliable and effective method of eliminating wood boring beetles, but because it is a costly, highly technical and hazardous process, you must not try to do this. Fumigation must be left to qualified pest control operators who are experienced in employing this technique.

Deathwatch Beetle adult (Anobiid)
(IX-9-A)

Deathwatch Beetle larva & damage
(IX-9-C)

Adult female Powderpost Beetle with extruded ovipositor and eggs
(WP-29)

Larvae of Powderpost Beetle; they often damage articles manufactured from hardwoods; i.e., flooring, ax handles, etc.
(WP-30)

Wharf Borer adult; larvae of these beetles cause damage to moist or wet wood in contact with water or soil.
(WP-21)

Old House Borer adult
(WP-25)

Furniture Beetle adult; these beetles are usually found damaging softwood structural timbers in the crawl space areas of homes.
(WP-33)

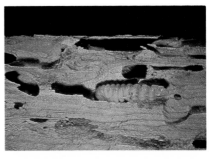

Old House Borer larva; note that frass has been removed from feeding tunnel.
(WP-27)

Southern Pine Sawyer adult; these insects are chiefly wood borers, but they bore in the inner-bark for some time before entering the wood.
(IBB-30)

Southern Pine Sawyer larvae; these beetles are pests of drying, bark-covered wood. Larvae feed into the wood and may cause damage in lumber cut from infected logs. The damage resembles that of the Old House Borer.
(WP-20)

Borers

COTTONWOOD TWIG BORER

The Cottonwood Twig Borer is a southern critter that feeds on the tender shoots of young trees causing the shoots to shrivel and break off. Texas host trees include cottonwoods, poplars and willows.

Adult beetles appear in mid-summer and after feeding briefly on the tender bark of the terminals descend to the base of the tree where the female deposits her eggs in small pits gnawed in the bark. Eggs hatch in about three weeks. The larvae then bore downward in the inner bark, entering a major root by fall. Larval feeding continues into the second year and in the process the larvae bore upward into the inner bark and wood of the host. The afflicted tree is severely weakened and may even be killed. Larvae transform into the non-feeding pupal stage and finally into an adult in the summer of the second year thus completing a two year life cycle.

Adult cottonwood beetles are rather striking in appearance. They are 1¼ to 1½ inches long and are black and white marked in a very vivid pattern.

Cottonwood Twig Borer = Order Coleoptera, Family Cerambycidae

Cottonwood Twig Borer adult
(CL-11-A)

PEACH TREE BORER

Peach Tree Borer adults will be in the field for only a short time, generally during July, August and September. During this peak period eggs are laid around the base of trees and hatch in about 10 days. Newly emerged larvae bore into bark near the hatching site and begin feeding in the cambium, that vital area just under the bark. Thick, gummy sap usually oozes from the entry holes. And now your tree is in big trouble. If the cambium layer is damaged extensively, the tree will die. Severely infested trees often can be identified by the dead or dying limbs.

It is not known how many larvae it takes to cause economic damage. A single feeding worm can devastate a young tree that is 1 to 2 inches in diameter. A tree 10 to 12 inches in diameter may safely harbor 1 or 2 larvae. Any more should be treated.

Peach tree borers begin life as eggs, then progress to larvae, then pupae, then finally to adult moths. The life cycle is completed in one year. Eggs laid in August and September mature into adults the next August and September.

Peach Tree Borer = Order Lepidoptera, Family Sesiidae

_____ **CONTROL CLUE** _____

This critter is a tough cookie and will require more than incidental spraying to control. Sevin and Lindane applied topside at the proper times, with a Thiodan drench at the base, can do the job, but I think an approved Dursban application is really the best. Make one application at the end of August. Do not contaminate the fruit; do not apply within 14 days of harvest. You really need to know much more than this. Get a control program from a responsible source. And don't waste your time with the old car-battery-hooked-to-tree cure. Won't work!

Peach Tree Borer female
(C-USDA-169)

Peach Tree Borer larva
(C-USDA-174)

SHOTHOLE BORERS

Shothole Borers are sometimes known as Fruit Tree Bark Beetles. An interesting "equation" might look like this: bark beetle = round-headed borer = shothole borer = peach tree borer = fruit tree bark beetle = peach twig borer = american plum borer = boring insects. (Don't try to prove Einstein's Theory with *this* equation.)

The point is this — several boring insects exist that will attack stone-fruit and other trees. Shothole borers are beetles; peach tree borers (not wood borers in the true sense), peach twig borers and American plum borers are small, clear-winged moths whose larvae are caterpillars which bore into the wood. Inconsistent genealogy notwithstanding, the result is dead fruit tree buds, twigs and limbs, and sometimes even dead fruit trees.

Female shothole borer beetles bore into the tree wood, creating tunnels in which they lay their eggs. The grubs that hatch bore into the inner wood, creating sawdust-filled burrows 2 to 4 inches long. The grubs pupate just under the bark, then emerge as adult beetles. The last generation of grubs spends the winter in the tunnels, emerging the following spring. Weakened, diseased, and dying trees and branches are most susceptible to borer infestation.

Shothole Borers = Order Coleoptera, Family Scolytidae

Shothole Borer evidence on bark
(C-USDA-173)

During the dormant season, remove and destroy infested branches. Apply a spray containing Lindane to the lower part of the trunk, wetting the bark thoroughly. Avoid spraying fruit and foliage. Severe borer infestation is usually the death knell for a tree. Sorry!

SQUASH VINE BORERS

The Squash Vine Borer is my personal Garden Enemy #1. (It's a cousin to the peachtree borer.) If I had my way, I'd hang this crook's picture in the Post Office. This bug does not fight fair. Look, I've just begun to write about it and already I have a headache. Excuse me, while I go get an aspirin.

Thanks. I believe I can make it now.

The adult moth is metallic green-black colored with hind wings fringed with black and orange hairs. It has similar colored markings over much of the abdomen. It is a day flier. Larvae (worms) are white, heavy-bodied, and over an inch long when full grown. And it's the larvae that do the dirty work. Here's their act: moths emerge in early summer and lay eggs on plant stems, normally close to ground level. Usually this occurs in April and May in southern Texas. On hatching, larvae bore into vines and that's the problem. That magnificent squash plant will wilt and soon die as the larvae tunnel up the hollow stem.

Squash Vine Borers = Order Lepidoptera, Family Sesiidae

Go after the adults with Sevin dust; begin treating as soon as you have a vine to treat. Keep the base of the plant well dusted. Pyrethrin spray will knock 'em down also, but there is no residual control. If you notice the wilt soon enough, find where the worm has entered the vine (it will be near ground level and will be covered with wet, tan sawdust-like mess). With a sharp knife, slit the stem vertically till you find the worm. Dig him out. Pack soil around the wound and it will likely heal. Remember, "an ounce of prevention . . ." or you also will have a headache — and no squash.

Squash Vine Borer adult
(Cole-TAEX)

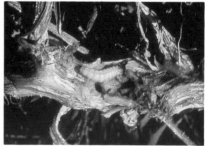

Squash Vine Borer larva & damage
(Cole-TAEX)

TREE BORERS

The predominant shade tree borers are the ROUNDHEADED and the FLATHEADED. Adult roundheaded borers are called LONG-HORNED BEETLES because of their long antennae. Adult flatheaded borers are called METALLIC WOOD-BORING BEETLES because of their iridescent metallic luster. The cylindrical, hard-shelled, long-horned beetle varies from ¼ inch to over 3 inches in length. Markings range from contrasting colored bands to spots or stripes. The antennae are usually from one-half to over one-and-a-half times the length of the body, depending on species. The metallic beetle usually is beautifully colored, is oval shaped and is ⅓ to 1 inch long.

The roundheaded group is more prolific and destructive, doing mischief by burrowing into the heartwood and leaving tunnels as large or larger than a pencil. Flatheaded borers damage or kill trees by mining beneath the bark or by tunneling into the heartwood and sapwood, leaving dead areas of bark with sap exuding.

Woodpeckers, expecially the Yellow Bellied Sapsuckers, cause damage to trees that is often attributed to borers, but the bird will likely make a row of holes in a more-or-less straight-line pattern circling the limb or trunk. The sapsucker revisits the tree many times, feeding on the sap accumulated in the holes it has drilled.

Flatheaded Beetles = Order Coleoptera, Family Buprestidae
Roundheaded Tree Borers = Order Coleoptera, Family Cerambycidae

CONTROL CLUE

I have bad news again! Once borers have entered a tree, control is usually fruitless. Nothing can be applied to the soil, or sprayed on or injected into the tree, that is effective. If only a tree or two is involved, you might try digging the borers out with a stiff wire, but don't bet on it. A *preventive* spray program using Lindane or Dursban is the only way to go.

Ponderosa Pine Bark Borer male
(CL-8-A VWR)

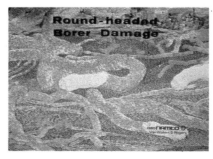

Round-headed Borer damage to wood
(CL-1-B VWR)

Fourspotted Longhorn Beetle adult
(CL-2-A VWR)

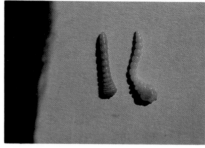

Roundheaded and Flatheaded Borer larva comparison
(Sandi Cole)

Metallic Wood Boring Beetle adult
(CM-1-A VWR)

Locust Borer adult
(C-USDA-143)

Stem Borer in cottonwood
(HI-22)

Boxelder Bug

Boxelder Bugs and other closely related insects are found throughout most of Texas, and they feed on several kinds of trees. In the fall they enter houses to find shelter for the winter, thus becoming a nuisance to many homeowners. They are plant feeders and do not feed on structures, food products, cloth, humans or pet animals; however, these bugs may stain curtains, paper and similar furnishings with fecal spots. They produce little or no odor when crushed.

Boxelder bugs are about ½ inch long and ⅓ inch wide. They are brownish-gray to black with distinctive red markings. Adults deposit eggs in cracks and crevices of tree bark in the spring, about the time buds begin to open. There may be two or more generations per year in Texas. They commonly hide in cracks and crevices in walls, door and window casings, and around foundations, as well as in tree holes and in large accumulations of debris. Adults are likely to come out of hiding and appear on light or white painted surfaces during warm days of winter or early spring.

Another insect, *Jadera haematoloma* (no common name), is often confused with the boxelder bug because it is so similar in appearance and habits. This insect is a beautiful bluish to smoky-black color with bright red eyes and stripes along the head. Apparently the development of this critter is similar to that of the boxelder bug.

Boxelder Bug = Order Hemiptera, Family Rhopalidae

CONTROL CLUE

Boxelder Bugs are not considered to be damaging to host plants. Although they feed by sucking plant juices, they are seldom abundant enough to harm trees. If insecticide treatment is necessary, it should be applied to young, exposed bugs found on host trees in the spring or early summer. Use an approved Dursban or Diazinon spray. Read the label.

Boxelder Bug adult
(HE-1-I VWR)

Boxelder Bug eggs on boxelder seed,
nymphs and adults
(HE-1-A VWR)

Cabbage Looper

Cabbage Loopers are voracious feeders and can strip foliage from an infested plant in short order. They enjoy a diversified menu, including cabbage, cauliflower, broccoli, brussel sprouts, lettuce, and occasionally beans, tomatoes and other crops. Larvae (worms) are light-green caterpillars with a few white or pale yellow stripes. They travel with a characteristic looping motion, hence the name Looper. Watching a looper travel tends to remind you of the hiccups.

Cooler temperatures slow their reproduction rate as it does with all other insects. There are continuous generations in the Lower Rio Grande Valley where normally temperatures are uniformly tepid. Mother Nature hits a lick for our side in that, when populations become crowded, a virus disease often strikes causing high larval mortality. Otherwise, I guess loopers would rule the world.

Cabbage Looper = Order Lepidoptera, Family Noctuidae

Cabbage Looper larva
(C-USDA-107)

Cabbageworm (Imported)

The Imported Cabbageworm's attack is similar to and easily confused with that of its kissin' cousin, the cabbage looper. Larvae are dark green caterpillars growing up to 2 inches long. They are frequently found right along with the looper, but there is one very obvious difference between the two. The imported cabbageworm travels without that looping motion . . . no hiccups. The menu and appetite of the two are about the same. They eat chosen vegetable leaves like crazy.

There is another most distasteful characteristic of cabbage cater-pillars. Their toilet habits are atrocious. Another form of damage by them is contamination of plants with greenish-brown excrement. And the bigger the worm, the bigger the . . . you know what.

Imported Cabbageworm = Order Lepidoptera, Family Pieridae

_____ CONTROL CLUE _____

Like for the cabbage looper. Sevin, Diazinon, Dipel or Thuricide. Read the label.

Imported Cabbageworm larva
(C-USDA-113)

_____ Cankerworms_____

Cankerworms are a major defoliator of broadleaf trees in east and central Texas. They are often called inch-worms or measuring worms because of their walking habit. They move by forming a loop with the central part of their body and they then extend their front to straighten out. Caterpillars (larvae) hatch from egg masses in the spring when trees are just reaching bud-break and new foliage is expanding. They grow to approximately one inch in length, are very slender and are extremely variable in color, usually being longitudinally striped, with green, brown and pale yellow colors predominant.

Cankerworms are a nuisance when they drop to the ground because they leave silk threads trailing from the trees. So, if you stand under an infected tree and look upward, be sure to keep your mouth closed. Many trees from Corpus Christi to San Antonio, north to Dallas and throughout east Texas were attacked in 1978 and 1979. Such outbreaks are not predictable and reach high levels with little warning.

Cankerworms = Order Lepidoptera, Family Geometridae

CONTROL CLUE

The surest cankerworm control is an early application of Orthene, Sevin, Dipel or Thuricide. Spraying is not recommended once these caterpillars are near full size. The damage will have already been done. The "early-bird" controls the cankerworm, so don't stand around every day just lookin' at 'em.

Parasitized larva of fall cankerworm
(HI-38)

Carpenterworm

Carpenterworms bore into the wood of living hardwood trees. The large winding tunnels in the sapwood and heartwood constructed by the larvae serve as an entrance for wood-rotting fungi and insects such as the carpenter ant. In cases of extreme infestation, the tree may be structurally weakened and subject to wind breakage. In Texas, oak species are preferred hosts, but black locust, maples, willows and fruit trees are also attacked.

Adult moths emerge in late April to early June, mating occurs, and then the females lay groups of eggs in bark crevices or wounds. Each mama lays 200 to 500 eggs during her one-week life span. After hatching, the larvae wander over the bark for a short time and then bore into the inner bark where they feed until they are half grown. They then bore into the sapwood and heartwood, returning occasionally to feed in the inner bark.

The larval period lasts from two to four years. Pupation usually occurs deep within the heartwood. Just prior to emergence, the pupa wiggles to the entrance hole where it remains slightly protruding until the adult moth emerges.

Carpenterworm = Order Lepidoptera, Family Cossidae

CONTROL CLUE

If you can catch this sucker in the larval stage when it's wandering over the bark, you've got a chance for effective control. Once it bores into the sapwood and heartwood, better you spend your time on a formula for retiring the national debt. Some chemicals which have a fumigating action have proved somewhat effective in controlling this insect in shade trees; no practical control has yet been found for forest trees. If you find an acceptable control, send it to me.

Carpenterworm larva in oak
(HI-16)

Carpenterworm moths, male & female
(HI-15)

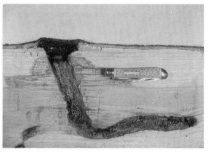

Carpenterworm gallery in oak made by two year old larva
(WP-37)

Caterpillars

SALTMARSH CATERPILLAR

We do have our share of caterpillars in Texas. And here's one that you hear a lot about. Saltmarsh Caterpillars eat everything — most garden crops, field crops, weeds, shrubs; they're certainly not particular.

Adults are white moths with black-freckled wings. Mama's wings are yellow on the underside. Papas have hind wings which are yellow above and below. These dudes are prolific breeders and make a lot of babies (worms). Larvae are up to 2 inches long when full grown and are covered with dense hairs ranging from yellowish to brown to nearly black in color. They migrate in a manner similar to armyworms, stripping foliage like a lawn mower.

Saltmarsh Caterpillar = Order Lepidoptera, Family Arctiidae

_____ **CONTROL CLUE** _____

A shot of Diazinon spray thoroughly applied should do the job.

Mating parents of Saltmarsh Caterpillars (Acraea Moths)
(LM-17-B VWR)

Saltmarsh Caterpillar larvae — "wooly-worms"
(Ento.-TAEX)

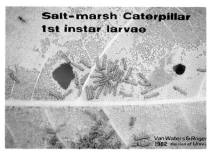

Saltmarsh Caterpillar 1st instar larvae on sycamore leaf
(LM-17-E VWR)

STINGING CATERPILLARS

Several kinds of stinging caterpillars occur in Texas, but the most common venomous larva is the fuzzy, white, tan or gray PUSS CATER-PILLAR, which is often incorrectly called an "asp." The IO MOTH LARVA, the SADDLEBACK CATERPILLAR and the HAG MOTH LAR-VA can hurt you also.

These caterpillars are the immature stages of various moths whose venom is conveyed through stiff, hollow, hair-like spines. These critters feed upon the foliage of many plants, including trees, shrubs and flowers. They rarely occur in numbers sufficient to damage plants, but they are important because of their medical effects.

A person's first symptom following contact with a puss caterpillar is an intense burning at the site of contact. The severity of the symptoms appears to be related to individual sensitivity, maturity of the larva, number of spines contacted, degree of pressure against the caterpillar, and site of the sting. In sensitive persons, lymph nodes under the arms or in the groin may swell and hurt; nausea, weakness and shock-like symptoms with severe headache may occur, usually within two hours after contact. Don't hesitate to see your physician. There is no really effective home first-aid treatment available. Prompt application of an ice pack and a baking soda poultice may help to reduce pain and prevent swelling. *Don't rub the affected area.*

Puss Caterpillar = Order Lepidoptera, Family Megalopygidae
Io Moth = Order Lepidoptera, Family Saturniidae
Saddleback Caterpillar = Order Lepidoptera, Famliy Limacodidae
Hag Moth = Order Lepidoptera, Family Limacodidae

CONTROL CLUE

Where stinging caterpillars present hazards to persons, such as around schools or residences, infested shrubs and trees may be sprayed or dusted with Dipel, Thuricide or Sevin. Remember, don't touch the critter.

Puss Caterpillar close up
(Ento.-TAEX)

Puss Caterpillar (white)
(John W. Norman)

Puss Caterpillar (tan)
(John W. Norman)

Io Caterpillar
(C-USDA-209)

Saddleback Caterpillar
(C-USDA-207)

Hag Moth Caterpillar
(C-USDA-211)

TENT CATERPILLARS

Four species of Tent Caterpillars are troublesome in Texas. EASTERN TENT CATERPILLARS and WESTERN TENT CATERPILLARS build large webs; SONORAN TENT CATERPILLARS build small tents; FOREST TENT CATERPILLARS build no tents at all. These species are closely related.

Tent caterpillars feed in groups and can defoliate a tree in short order. Mamas are moths. Larvae hatch in early spring at about the time host plants leaf out. The tents of the Eastern and the Western caterpillars are most often found in a crotch of small limbs at the trunk of the tree and these tents enlarge as the colony grows. Larvae move from these refuges to feed on leaves, so damage can be found for some distance around the web.

Tent Caterpillars are rather like people in the sense they all do about the same things, but differ only in appearance. Larvae (worms) are attractively colored, are about one and a half inches long and have a few long white hairs on their bodies, mostly along the sides. The Eastern has a solid white line down the center of its back; the Forest has a row of "key-hole" shaped white marks, one on each body part; the Western always has a series of white dashes down the middle of its back; the Sonoran has no white marks, but rather sports a series of yellow dashes and blue spots on each segment.

Tent Caterpillars = Order Lepidoptera, Family Lasiocampidae

Eastern Tent Caterpillars — solid line;
Forest Tent Caterpillar — foot prints
(TFS-BILLINGS)

Tent mass of Eastern Tent caterpillar
(HI-30)

Forest Tent Caterpillar adult moth on oak
leaf   (LM-4-F VWR)

Forest Tent Caterpillar larva on oak leaf
close up   (LM-4-B VWR)

Forest Tent Caterpillar after shower and
shave   (TFS-Billings)

Eastern Tent Caterpillar larva
(C-USDA-156)

Western Tent Caterpillars and tent
(ORTHO-48)

CONTROL CLUE

There is only one generation of tent caterpillars a year. If they have been allowed to feed and have completed their development, it is useless to spray, so hit 'em early with Sevin, Thuricide, Diazinon, Malathion or Methoxychlor.

VARIABLE OAK LEAF CATERPILLAR

The Variable Oak Leaf Caterpillar periodically defoliates hardwood trees in eastern Texas. The larvae feed primarily on oaks, but will also feed on beech, basswood, birch and elm. Young larvae skeletonize the leaf, while older larvae devour the entire leaf except for the primary veins. While infestations usually subside before many trees are killed, heavy defoliation reduces the tree's vigor and growth.

The variable oak leaf caterpillar overwinters as a non-feeding larva in a cocoon on the ground. It pupates and emerges as a moth the following spring. The female moth, gray in color and about 1¾ inches long, lays about 500 eggs singly on the leaves of host trees. The hatched larvae feed on foliage for five or six weeks, drop to the ground to pupate and emerge as adults in mid-summer. Larvae hatching from eggs laid by the second generation of moths defoliate the trees for the second time during late summer. By late October the mature larvae of the second generation have dropped to the ground to overwinter.

The full grown larva is approximately 1½ inches long. The body is usually yellow-green with a narrow, white stripe down the center of the back bordered by wider dark bands. The head is usually amber-brown with curved diagonal white and black bands. Colors will vary among individuals.

Variable Oak Leaf Caterpillar = Order Lepidoptera, Family Notodontidae

CONTROL CLUE

Outbreaks of the variable oak leaf caterpillar may be severe, but usually subside before tree death occurs. Although mice and predaceous beetles feed on the resting larvae and pupae, other predators and parasites generally are not effective in controlling rising populations of this critter. No chemical is currently registered for control of this insect; however, Sevin, Thuricide or Methoxychlor have been effective and safe in controlling related caterpillars.

Variable Oak Leaf Caterpillar adult
(HI-41)

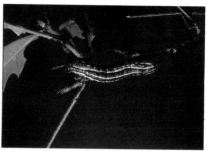

Variable Oak Leaf Caterpillar larva
(HI-46)

Variable Oak Leaf Caterpillar larvae and
damage
(HI-45)

WALNUT CATERPILLAR

Walnut Caterpillars are a serious threat to pecan, hickory and walnut trees. Oak, willow, honey locust, and certain woody shrubs can be endangered also. In Texas, at least two generations develop each year, with the second being more prolific, thereby causing more damage. In years past, almost all of the native pecan trees in certain areas of the state have been defoliated by these critters. They are strictly bad news.

The adult is a moth with a two-inch wing span. It is brown and tan with a dark region on the body behind the head with wavy, dark lines across the front wings. The female will deposit about 300 eggs on the underside of a leaf. Caterpillars (larvae) hatch in about nine days and live together in a group, growing up to two inches in length. They do not make webs. They are reddish-brown to black with white markings and have long, fuzzy, white hairs. Larvae characteristically arch their heads and tails in a defensive posture when disturbed.

Walnut Caterpillar = Order Lepidoptera, Family Notodontidae

CONTROL CLUE

To avoid as much damage as possible, spray these caterpillars when they are still young and small. Thoroughly wet leaves, twigs, limbs and tree trunks with Sevin, or Orthene, or Diazinon or Thuricide, the choice depending upon variety of tree. Remember, read the label on the container and follow directions. Large trees may require high-pressure spraying for effective control. And hurry.

Disturbed Walnut Caterpillars; note arched heads and tails
(TFS-Pase)

Walnut Caterpillar orgy on a pecan tree
(Ento.-TAEX)

Walnut Caterpillars (just washed their hair and can't do a thing with it)
(Ento.-TAEX)

Senior prom — Lufkin, Texas; Walnut Caterpillar Hi, class of '79
(TFS-Billings)

_____ Catfacing Insects _____

Catfacing is a disfiguration of fruit mostly caused by a group of sucking insects and some chewing insects. It is a major problem in Texas. This malformation is caused by an attack on the fruit when it is in an early, formative stage. These insects, various stinkbugs and plant bugs, penetrate the fruit with their sucking mouthparts.

Early in the season at the pink bud stage, Lygus Bugs attack peaches. These insects are about 1/5 inch long and range in color from shades of brown to tan to nearly black. They damage buds and blossoms as well as small fruit. Several species of stink bugs attack peaches and plums in Texas. Their damage may occur early in the fruit's development, but the blemishes remain on the final product. Damage caused by the feeding of the plum curculio can also cause catfacing in most fruits, particularly in peaches, plums and apples.

Hail or cold weather damage to tender blooms may also cause catfacing. Tomato catfacing can be caused by extreme heat (above 85 degrees Fahrenheit) or by cold (below 55 degrees Fahrenheit) or by drought.

Stinkbugs = Order Hemiptera, Family Pentatomidae
Plantbugs = Order Hemiptera, Family Miridae

Mother Nature will determine the temperatures, but we can handle those insects. Spray with Malathion 50% E.C. (emulsified concentrate-liquid) when ¾ of the bloom petals have fallen. Clean up weeds and plant debris in the fall to eliminate hibernation locations for overwintering bugs.

Tarnished Plant Bug adult   (HE-12-B VWR)

Harlequin & Green Stink bugs   (HE-11-A VWR)

Lygus Bug adult   (HE-9-A VWR)

Catfacing plant bug injury on peach
(C-USDA-176)

Mirid Plant Bug on cucumber leaf
(HE-13-A VWR)

Centipedes and Millipedes

Centipedes and Millipedes are usually considered nuisances rather than destructive pests. Centipedes pose an occasional threat to man because they are equipped with poison glands and they will bite. The poison usually produces only a moderate reaction similar to that of a bee sting, but if you are bitten and have a significant reaction, don't hesitate to see your physician.

Most centipedes feed upon small creatures such as insects. With powerful jaws, which are located immediately behind their head, they grasp their prey and kill it by injecting venom. They prefer moist, protected habitats such as under stones, rotted logs, leaves, bark or in compost piles.

Millipedes are not poisonous, but many species have repugnatorial glands capable of producing irritating fluids which may produce allergenic reactions in sensitive individuals. A few millipede species are capable of squirting these fluids over a distance of several inches. Persons handling millipedes will notice a lingering odor on their hands and the fluid can be dangerous to the eyes, so don't fool around with them. Millipedes feed primarily on decaying organic matter, but may attack roots and leaves of seedling plants. Their favorite abode is the greenhouse.

Centipedes = Class Chilopoda, Order Scutigeromorpha, Family Scutigeridae
Millipedes = Class Diplopoda, Order Chordeumida, Family Lysiopetalidae

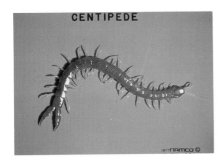

Centipede
(IV-30-A VWR)

Millipedes: one walking; one coiled
(IV-31-A VWR)

_____ Chiggers_____

Common Chiggers, also known as "redbugs," attach themselves to tender skin areas to feed. These very tiny pests belie their size in the amount of torment they cause humans. Their bites cause intense itching and small reddish welts on the skin. Most persons cannot see chiggers without a magnifying glass, so the bites may be the only indication that the bugs have infested a given area.

Chiggers are related to ticks and spiders. The young chigger (larva) which hatches in the spring is the real troublemaker. It is a parasite that feeds on man and animals. This wee, bright red larva can scarcely be seen as it scurries along the skin surface seeking an attachment site. When it finds one, such as a skin pore or hair follicle, it attaches its mouthparts to the spot. On people, the chigger prefers places where clothing fits tightly or where the flesh is thin or wrinkled.

Contrary to common belief, the chigger does not burrow into the skin or suck blood. Instead, it injects a digestive juice that disintegrates skin cells so they can be used as food. Affected skin tissue becomes red and swollen. It may completely envelop the feeding chigger, making it appear to be burrowing into the skin. The bite itches intensely and may continue itching for several days even after the chigger is killed or drops off.

Chiggers = Class Arachnida, Order Acari, Family Trombiculidae

Chigger — not magnified. (Don't gripe! I told you they were little.)

Chinch Bugs

Chinch Bugs can damage home lawns severely, particularly during hot, dry weather. Infested lawns first turn yellow, then turn brown and then die. In Texas, adult chinch bugs are inactive during the winter, and even in the spring cause little damage, but in the summer and early fall . . . look out.

Chinch bug infestations can be accurately diagnosed only if the insects are found. When damage is severe and bugs are plentiful, they usually can be found simply by parting the grass and carefully observing the soil surface. For early detection, chinch bugs can be found by pressing one end of an open-ended coffee can about 2 or 3 inches into the soil at the edge of the beginning-to-yellow grass. Fill the can with water and keep it nearly full for about 5 minutes. Any chinch bugs present will float to the surface. And then you know the awful truth.

Adult chinch bugs are about 1/6 to 1/5 of an inch long. They have black bodies with white wings. Each wing bears a distinctive triangular black mark. After hatching, the wingless nymphs are yellow, but soon turn red and develop a light-colored band across their backs. Before the last molt, nymphs are black or brownish-black with a white spot and two small wing pads on their backs.

Chinch Bugs = Order Hemiptera, Family Lygaeidae

CONTROL CLUE

Best control results when the entire lawn is treated, and watering the lawn before insecticide application will aid penetration. Use Dursban granules and apply with your fertilizer spreader according to directions on the bag.

Chinch Bug adult & nymph on grass
(HE-2-A VWR)

Chinch Bugs, males & females
(HE-2-D VWR)

Cicadas

Cicadas are frequently misidentified as locusts. Even today their appearance arouses fear of crop destruction. American Indians once thought these curious creatures had an evil significance. These critters are more frequently heard than seen, being responsible for that screeching cacophony during hot, late-summer afternoons. Only the males sing.

The females' egg-laying activity can severely damage or destroy twigs and small branches of vines, shrubs and trees. Eggs hatch in 6 to 7 weeks. The resulting ant-like nymphs drop to the ground and burrow through the soil to find roots from which they suck juices. Under some conditions, the nymphs may build mud cones or chimneys 3 or 4 inches high from which they emerge from underground. The emergence hole in the mud cone (or in the soil if no cone is constructed), is approximately ½ inch in diameter.

Adult emergence from the nymphal skin takes place at night on trees, posts or buildings. This light brown, brittle, beetle-shaped shell that you are so reluctant to touch will hang on a tree trunk for months if left undisturbed. It's harmless.

Cicadas = Order Homoptera, Family Cicadidae

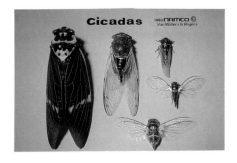

Five different species of Cicadas; size range — ½ to 3 inches
(HC-1-A VWR)

_____ Cockroaches _____

Approximately 3500 species of Cockroaches exist in the world today; in Texas only five are really troublesome: American, Smokybrown, German, Oriental and Brown-banded. Most cockroaches are tropical or sub-tropical in origin and generally live outdoors, but some are perfectly comfortable living with man.

Although generally identified with a filthy environment, cockroaches will at times infest even the most sanitary and well organized homes and buildings. Most are nocturnal and appear during daylight only when disturbed or where there is a heavy infestation. Roaches enter buildings in infested boxes, seasoned firewood, grocery bags, beverage cartons or other containers. They also enter of their own volition around loose-fitting doors and windows, where electrical lines or water pipes pass through walls, or through sewer lines.

Cockroaches are some of the oldest insects, as indicated by fossil remains dated to 200,000,000 years ago. We contend with a formidable adversary in attempting to eliminate the cockroach, but there's no choice — either fight him or give him a deed to the house!

Cockroaches = Order Orthroptera, Family Blattidae (Oriental Cockroach)
Family Blattellidae (German Cockroach)

CONTROL CLUE

Study each cockroach problem and use control measures in accordance with the location, extent and nature of the infestation. Hit 'em with everything available, including your shoe; put out roach baits; use contact insecticides and residual sprays (Dursban is good). I remember the sage advice of my friend, Bill Spitz, President of Big State Pest Control: "S.O.S. — Seek Out the Source." Employing a capable professional exterminator will be your easiest solution.

If you want a surefire check for roaches in your home, invite the snobbiest people you know over for dinner. If there is a roach in the house, I promise it will run across the floor that evening. And if you see a roach, don't scream R-O-A-C-H. Quietly say "My, what is that!" Your husband should reply, "Oh, just a water bug." Don't lose your cool!

American Cockroach
(C-USDA-194)

Smoky Brown Cockroach
(II-2-J VWR)

Brown Banded Cockroach
(C-USDA-197)

German Cockroach
(C-USDA-196)

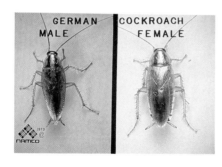

Oriental Cockroach
(C-USDA-195)

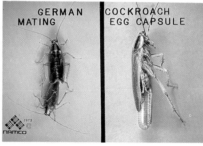

German Cockroaches mating & female with egg capsule
(II-5-A VWR)

German Cockroach adult male/female comparison
(II-5-G VWR)

American Cockroach & house mouse droppings comparison
(II-1-K VWR)

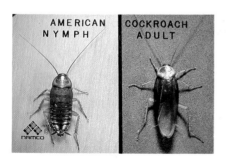

American Cockroach nymph & adult
(II-1-G VWR)

Codling Moth

Codling Moths are one of the two varieties of fruit moths that attack deciduous fruits in Texas. Their behavior is very similar to that of the Oriental Fruit Moth, but they attack the fruit of apples and pears more directly, causing the fruit to drop early in the season. Larvae (worms) tunnel into the fruit and feed for several weeks. The entry holes are rather obvious because of the mass of brown, sawdust-like material surrounding them. Once invaded the fruit is ruined.

Codling Moth = Order Lepidoptera, Family Olethreutidae

CONTROL CLUE

Once the fruit has been penetrated by the larvae, it's too late for control measures. Employ the same procedures as for the Oriental Fruit Moth. Remember, the early-bird gets the worm.

Codling Moth adult
(C-USDA-172)

Corn Earworm

The Corn Earworm, alias Tomato Fruitworm, alias Cotton Bollworm, alias Sorghum Headworm . . . all the same critter! The reason the corn earworm doesn't know he's a corn earworm is simply the fact that he doesn't read these insect books we humans write. This caterpillar is the most serious pest of corn, but also it is most devastating to tomatoes and often to cabbage, beans and peppers. And this maniac is not above slipping over to your rose bush for yet another course on its menu.

Larvae (worms) may reach an inch and a half in length and may vary in color from green to pink to almost black. Alternating longitudinal dark and light stripes mark this worm's body, but colors are so variable that such characteristics are not dependable for identification. You will best know him by the fruit he eats. And can he eat. Remember, this caterpillar tunnels into fruit and/or feeds on leaves. Good luck with this one.

Corn Earworm = Order Lepidoptera, Family Noctuidae

CONTROL CLUE

Sevin is about as good a control as any, but remember to hit him when he's little and damage is first noticed. Remember, no control . . . no tomatoes.

Corn Earworm with corn damage
(C-USDA-32)

Sorghum Headworm
(C-USDA-88)

Tomato Fruitworm
(C-USDA-92)

Cotton Bollworm
(C-USDA-73)

Cowpea Curculio

The Cowpea Curculio might have a fancy name, but it's still just a cotton-pickin' weevil. Adults are about ¼ inch long, are black, and possess that distinctive weevil snout. The larvae (grubs) are about ¼ inch long and are legless, white and C-shaped. Damage is done primarily by the chewing grub.

These critters like beans . . . string beans, lima beans, wild beans, southern peas or cowpeas, but if times get tough, they'll even go after cotton seedlings. Here's how they do their thing. Mama curculio pierces a developing bean pod with her long snout and deposits eggs. Hatching occurs in about three days and grub-baby begins eating beans. The entire life cycle, from egg to adult, may be completed in 30 days. And once they start this jazz, kiss your beans goodbye.

Cowpea Curculio = Order Coleoptera, Family Curculionidae

Cowpea Curculio attacking bean pod
(C-USDA-116)

_____ Crickets_____

Of the several species of crickets, House Crickets and Field Crickets
are the most common and are the most troublesome. Both of these
crickets have antennae that are longer than the body, and both are good
jumpers and fliers. Males have two appendages extending from the tip
of the abdomen; females have three. Males make the noise.

FIELD CRICKETS prefer to live and breed outdoors where they
feed on several kinds of plants. Occasionally, they invade homes in
search of warm hiding places, but will not breed or establish a permanent infestation indoors.

HOUSE CRICKETS commonly breed outdoors, but unlike field
crickets, can live and breed indefinitely indoors. The fabled Jiminy
Cricket notwithstanding, these rascals are pests. Outdoors, they damage

garden plants. Indoors, they can damage woolens, cottons, silks, synthetic fabrics, furs and carpeting. Clothes stained with perspiration are particularly attractive. They also feed on foods, leather and rubber products.

CAMEL CRICKETS, also called Stone or Cave Crickets are not true crickets, are wingless, and are otherwise distinctive in appearance because of their arched backs. They are active at night, are not attracted to light and have no "song."

JERUSALEM CRICKETS are sometimes called "children of the earth." They have large, round, naked heads with two bead-like black eyes that may give them a fancied resemblance to a miniature child. They burrow into loose soil, are active at night and are seldom seen. They are useful predators, feeding on other insects and spiders. Although fierce looking, Jerusalem Crickets are harmless, but their powerful mandibles could inflict a minor, nonvenomous wound if handled carelessly.

Crickets = Order Orthroptera, Family Gryllidae

CONTROL CLUE

INDOORS: Apply a residual spray of Dursban, or find the critter at night with a flashlight and bop him with a shoe, unless you are superstitious about killing crickets. (First-degree cricket-murder is supposed to bring bad luck, but if you apply an insecticide for roaches and Brother Cricket happens to wander into it and dies, then it's either accidental death or suicide and you're off the hook. There's not a jury in the world that would convict you.) OUTDOORS: Dursban or Diazinon properly applied should do the job. Read the label.

House Cricket — female & male
(IV-20-A VWR)

Field Cricket — adult
(IV-19-A VWR)

Camel Cricket adult
(C-USDA-204)

Jerusalem Cricket adult
(IV-21-A VWR)

Snowy Tree Cricket on tobacco
(C-USDA-63)

MOLE CRICKETS

Mole Crickets are about one and a half inches long, are golden to chocolate brown in color and are covered with fine velvety hairs. They are endowed with strong front legs equipped for digging. And dig they do. Their small winding burrows of loosened soil reveal their presence. They are nocturnal feeders and may tunnel as much as 10 to 20 feet a night.

Those meandering surface ridges that suddenly appear on your garden soil indicate the presence of mole crickets. The next symptom will be some dead or damaged plants. Two common mole cricket varieties attack Texas vegetables — the Northern Mole Cricket and the Southern Mole Cricket. Almost all vegetable crops may be damaged by these pests.

Mole Cricketts = Order Orthroptera, Family Gryllotalpidae

_____ **CONTROL CLUE** _____

Diazinon granules scratched into the soil in proximity to his tunnels should do the trick.

Mole Cricket
(Ortho-38)

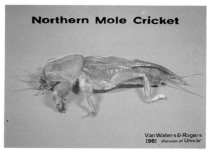

Northern Mole Cricket adult
(IV-22-A VWR)

_____ Curculio (Plum) _____

Plum Curculios are the reason you have wormy plums and peaches every year. Some damage is caused by adult weevils feeding, but the worst damage is caused by the larvae. Here's how curculios do their number on your fruit crop. When mama weevil gets the maternal urge, she will pierce a plum with that long snout and oviposit (lay eggs). At this point you can kiss that plum goodbye. When the eggs hatch, larvae (worms) develop and begin tunneling through the fruit, eating all the way. Left to their own devices, these grubs will eat more plums and peaches than you will.

Infected fruit usually falls to the ground, maturity notwithstanding. After the larvae complete their development, they pupate in the ground and become adults. These adults return top-side to repeat the carnage. Two or three generations of this pest can be produced in a single season, and they will begin their mischief early, even on a match-head size plum. One of the key things to look for is a characteristic C-shaped or crescent-like cut on the fruit. This scar is formed by the female during the egg-laying process. If you see a "C," you have a curculio problem. Stout-hearted gardeners fight back, and here's how you do it.

Plum Curculio = Order Coleoptera, Family Curculionidae

CONTROL CLUE

Here's another case of *when* you spray being as important as *what* you spray. Use Malathion 50% E.C. (emulsified concentrate-liquid). Follow the directions on the label. When ¾ of the bloom petals have fallen, *spray*. This must be repeated every 10 to 14 days throughout the fruit growing season. The timing of the first spray will be critical. If you miss it, the remainder of your spray program will be mostly in vain. Once the fruit is infested, there is no way to kill the grubs inside the fruit. Oh, the price we mortals must pay for plums and peaches — without worms!

Plum Curculio and plum damage
(C-USDA-171)

Plum Curculio larva in plum
(C-USDA-175)

Cutworms

Here are some guys who can make vegetable gardeners cry. SUBTERRANEAN CUTWORMS feed almost entirely below the soil surface, eating roots and underground stems — Pale Western Cutworms are important members of this group. TUNNEL DWELLERS form and live in tunnels — stars of this group are Black Cutworms which will cut tender plants at the soil surface, pull them into their tunnels and eat 'em. Army Cutworms are night SURFACE-FEEDERS who spend the daylight hours hiding under soil, mulch or trash. Granulated Cutworms are other surface feeders that do serious mischief to garden vegetables in central and south Texas. CLIMBING CUTWORMS, the athletes of the group, feed during the night and hide under boards and rocks during the day — Variegated Cutworms are good examples.

Often times, the cutworm may be found during daylight hours (sleeping off his gluttony) curled up just below soil level in proximity to his latest conquest. If you find one, bust him with a rock, stick or any thing handy. (I have found he pops readily if squeezed between thumb and forefinger, but don't let your wife see you do this!)

Cutworms = Order Lepidoptera, Family Noctuidae

CONTROL CLUE

A more civilized and perhaps more effective control would be applying a cutworm bait. Sprinkle the granules at the base of each seedling when you plant it. If the distasteful method (squeezing) or the civilized method (bait) is not an appealing remedy, then you might want to try the old-time self-rising flour remedy. Sprinkle the flour, as you would the granules, around the seedling when you plant it. Mr. Cutworm does like that flour. He will come in the night and eat his fill. The next morning you will find a king-size cutworm, dead as a doornail. Caution! Use *self-rising* flour. He will thrive on regular flour and in a couple of days you will have raised the biggest, meanest cutworm "Go-rilla" in the neighborhood.

Black Cutworm and damage to corn
seedling
(C-USDA-97)

Climbing Cutworm and damage
(C-USDA-53)

Variegated Cutworms in the morgue
(Ento.-TAEX)

_____ Diamondback Moth _____

Here's a little critter that causes some pretty big trouble. The adults are grayish moths which are about ⅓ inch long. Mom and pop aren't so bad, but again it's the blankety-blank kids (larvae) that do the damage. These little devils, which rarely exceed ⅓ inch in length, are pale yellowish-green caterpillars with fine, erect, scattered black hairs over their bodies. And they wiggle like the dickens if disturbed. They feed on the underside of leaves, leaving shot-hole type damage and you go nuts trying to decide what's eatin' your plant. Usually the outer leaves are attacked. What plants are preferred? Cabbage, cauliflower, broccoli and stuff like that.

Diamondback Moth = Order Lepidoptera, Family Plutellidae

_____ **CONTROL CLUE** _____

Use Sevin, Dipel or Thuricide as you would for any other garden caterpillar, but be sure to spray or dust the underside of the leaves. You should always do that anyway.

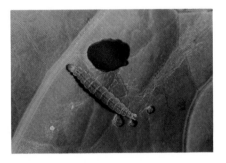

Diamond Back Moth larva with damage to
leaf
(C-USDA-112)

_____ Earthworms _____

Earthworms in your garden? You should be so lucky! As they tunnel through the ground, earthworms ingest soil and in the process digest any organic matter in it. They usually deposit the soil that passes through their bodies as crumbly mounds (castings) on the soil surface. Although they are most numerous within the top 6 inches of soil, earthworms may tunnel as much as 6 feet down, bringing up deep layers of soil to the surface. These tunnels help to aerate and loosen the soil, which in turn improves soil drainage and tilth and facilitates plant root growth.

Earthworm activity in your ground is beneficial and should be encouraged. Gardeners who consider them a nuisance don't understand Mother Nature. When I till my garden between seasons and uncover a bumper crop of earthworms, I know several things: my garden soil is healthy; humus content is good; moisture level is adequate; there is no active insecticide prevalent; and I'm on target for a successful next garden. I would rather suffer nematodes and grub worms than fumigate and destroy these benefactors. I love earthworms . . . and so should you.

Earthworms = Phylum Annelida

_____ **BENEFICIAL**_____

Here's ol' good-buddy, but you can't tell
whether he's coming or going
(Mike Mitchell)

_____ Earwigs _____

Earwigs are hard, flattened, reddish-brown insects, up to 1 inch long. They are usually found in the garden, where they feed on mosses, decaying organic matter, vegetation and other insects, but they will also invade homes. They usually enter through cracks or openings in the foundation, or through doors, windows or other places. They are more apparent during hot, dry spells. In the home, they do not damage household furnishings, but their presence is annoying. From their abdominal glands, earwigs exude a liquid that has a disagreeable, tar-like odor. They are often found in dark, secluded places such as in pantries, closets, drawers and even in bedding.

Earwigs are only minor pests in the garden unless populations are high. They feed at night and hide under stones, debris and bark in the daytime. Adult earwigs lay eggs in the soil in late winter to early spring. The young that hatch may feed on green shoots and eat holes in leaves, and as they mature, may feed on blossoms and mature fruit.

Earwigs become beneficial when they feed on other insect larvae and snails. Those frightening, forcepslike pincers that extend from the back end are used as offensive and defensive weapons with which they catch insects; earwigs are also capable of giving you a good pinch if you mess with them, but the legend of their creeping into ears of sleeping persons is untrue.

Striped Earwig = Order Dermaptera, Family Labiduridae

_____ **CONTROL CLUE** _____

Because earwigs typically cause only minor damage in the vegetable garden, insecticide sprays are seldom needed. In the house, use an approved Diazinon spray; or bop him with a shoe if you don't mind that disagreeable odor.

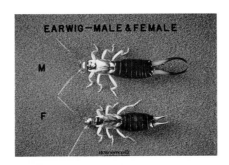

Earwigs; male and female
(IV-18-B VWR)

Fabric Pests

CLOTHES MOTHS AND CARPET BEETLES

Two types of clothes moths are common to Texas — CASEMAK-ING CLOTHES MOTH and WEBBING CLOTHES MOTH. Larvae of the casemaking clothes moth feed from a silken case which is dragged over the surface of their food; webbing clothes moth larvae feed within silken burrows which they spin over the fabric surface. Adult clothes moths are winged, buff-colored, and about ½ inch long. They are weak fliers and avoid lighted areas. The larvae do the damage. They commonly feed on wool, feathers, fur, hair, upholstered furniture, leather, fish meals, milk powders, lint, dust, paper, or even synthetic materials which may be soiled with oils; in other words, they will eat everything but the kitchen sink.

Adult CARPET BEETLES do not eat fabrics, but their presence in the springtime inside and outside the home should alert the homeowners. They may be seen crawling up walls of infested homes and also may be found congregating on window sills. BLACK CARPET BEETLES are black with brown legs; COMMON CARPET BEETLES, FURNITURE CARPET BEETLES and VARIED CARPET BEETLES are mottled white, yellow, red and black. The body is usually covered with scales. Larvae of the black carpet beetle are carrot-shaped, are covered with brown bristles, with a tuft of long brown hair protruding from the end of the body. Larvae of the other species are short, stocky and are covered with brown and black bristles. Remember, the larvae do the damage by feeding on animal products such as wool, silk, hair, fur or feathers. They may feed for 9 months to 3 years, destroying carpets, felt padding, mohair cushions, furs and many other household items. These larvae can move from one type of food to another.

Casemaking and Webbing Clothes Moths = Order Coleoptera, Family Dermestidae
Carpet Beetle = Order Coleoptera, Family Dermestidae

CONTROL CLUE

The key to economic control of Clothes Moths and Carpet Beetles is good housekeeping. All furs, woolens, etc. should be sunned, brushed and dry cleaned periodically. Sweep or vacuum regularly to remove lint, hair and dust from floors, shelves and drawers where adult clothes moths and carpet beetles may lay eggs and establish infestations. Give close attention to rugs, carpets, draperies, furniture cushions, closet corners, cracks, baseboards, moldings and other hard-to-reach places. Look for sources of infestation around the home such as old clothing, woolen scraps and yarn, furs, feather pillows and piano felts. It is vitally important that all cloth goods be dry-cleaned or washed, pressed with a hot iron, and brushed prior to storage in order to rid them of insects.

Do *not apply sprays to furs;* cold storage is the most practical method of damage prevention for furs. Cedar-lined closets and cedar chests have limited value in fabric pest protection unless other measures are used. One pound of napthalene flakes or balls, or paradichlorobenzene (PDB) crystals per 100 cubic feet of closet space will provide adequate protection, but the area and contents will smell like "moth balls."

Rugs and carpet surfaces, especially around the edges and under heavy furniture, should not be ignored. Piano felt pads are best treated by a piano technician; upholstered furniture and pillows may require fumigation by a professional pest control operator because surface sprays will not control fabric pests inside the stuffing. Surface applications of Malathion, Diazinon or Methoxychlor should be made as coarse sprays along the edges of wall-to-wall carpeting in closets, corners, cracks, baseboards, moldings and other secluded areas. These treatments are particularly important for carpet beetles since they commonly live in lint and debris in these areas.

Are you sure you can do all this? Most good pest control firms provide dependable service for controlling Clothes Moths and Carpet Beetles; why don't you call one and get this job done right!

Casemaking Clothes Moths   (C-USDA-17)

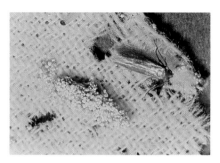

Webbing Clothes Moth
(C-USDA-04)

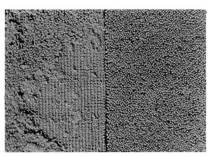

Webbing Clothes Moth damage to carpet
(C-USDA-05)

Black Carpet Beetles
(C-USDA-01)

Black Carpet Beetle damage to shirt
(C-USDA-02)

Furniture Carpet Beetles
(C-USDA-08)

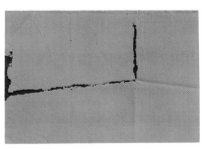

Furniture Carpet Beetle damage
(C-USDA-09)

Fleas

Fleas are found worldwide. There are an estimated 1600 species. The most common fleas encountered by Texans are Cat Fleas. Others, such as Dog Fleas, Human Fleas, Oriental Rat Fleas and numerous other rodent fleas are encountered only occasionally by the homeowner. Female fleas lay eggs shortly after a blood meal. Because fleas have the ability to survive for many months without food, they can remain in suitable areas for long periods of time waiting for dinner, which usually is your cat, dog, or you. Occasionally eggs are laid while the female flea is on the host, but the eggs normally fall off. This is how homes become infested. Cat and dog fleas may also be found on humans, rabbits, squirrels, rats and birds. Human fleas also attack swine, goats, cats, dogs, rats, coyotes and skunks. Like, one big happy family!

Cat Flea = Order Siphonaptera, Family Pulicidae

CONTROL CLUE

Successful flea control must include treatment of the infested animals and also thorough treatment of the entire premises — indoors and outdoors. The effectiveness of pesticides is directly related to the thoroughness of applications. A number of insecticides are approved for specific uses. I'll give you a menu, but remember there are others. For the pet, ask your veterinarian; indoors, Bendiocarb; for the yard, Dursban granules. Maintain applications for 7 to 10 days to break the egg-hatch cycle. Or, call a qualified pest control service and be done with it.

Flea — magnified
(Ortho-25)

Cat Flea/Squirrel Flea comparison
(III-1-L VWR)

Fleahoppers

When you hear conversations about Fleahoppers in Texas, it is likely to be among cotton farmers complaining about the damage these pests are responsible for in the cotton patch. The one that you are likely to be concerned with is the Garden Fleahopper. Male adults are winged and so are the females, but her wings are shorter. Both are nearly black in color. Fleahoppers have large hind legs that enable them to hop actively, which they do. Nymphs look like adults, except they are smaller, are greenish in color, and are wingless. Fleahoppers attack southern peas, other garden crops, and weeds. They are sap-suckers, making small discolored areas on the foliage. They can kill leaves, thereby injuring plants seriously.

Fleahoppers = Order Hemiptera, Family Miridae

CONTROL CLUE

Spray or dust plants with Malathion or Sevin.

Cotton Fleahopper
(Charles L. Cole)

Flies

There are a great many species of Flies; some are dangerous to man as carriers of diseases and some are destructive to crops. But some are useful scavengers that clean up dead animals and plant wastes; others are insect destroyers — either predators like the Syrphid Flies or parasites like the Tachinid Flies that live in or on harmful insects. The larvae, called maggots, are footless, grub-like creatures that are usually soft, white or yellowish, with a reduced head.

HOUSEFLIES are common throughout the world. In addition to being annoying, they can spread a number of serious human diseases and parasites such as diarrhea, dysentery, typhoid, cholera, intestinal worms and Salmonella bacteria. Flies feed on and lay their eggs in decaying organic matter; the eggs can hatch within 12 hours under ideal conditions. The creamy-white maggots burrow into and feed on the decaying material, pupate and then emerge as adult flies. The entire life cycle may be completed within 14 days. Several other fly species may infest the home, include FACE FLIES and LITTLE HOUSE FLIES.

FRUIT FLIES do not constitute a serious health menace, but can be annoying where fruit, vegetables or garbage is allowed to rot and ferment. The adult flies lay their eggs in the decaying fruits or vegetables. The eggs hatch in a few days, and the tiny maggots feed on yeasts growing in the decaying food. These tiny (up to 1/6-inch) VINEGAR FLIES are yellowish-brown, are clear-winged, and fly in a slow hovering manner. Other fruit flies of extreme economic importance in agriculture are MEXICAN FRUIT FLIES and MEDITERRANEAN FRUIT FLIES. The Mexfly and the Medfly are subject to quarantine laws and continuous detection programs are monitored by state and federal agencies in order to keep these "biological-atomic-bombs" under control.

HORSEFLIES and DEERFLIES attack humans and domestic animals in rural and suburban areas. Female flies deposit their eggs in still pools of water, in moist soil or on vegetation. The larvae feed on decaying vegetation or on other insects and pupate in damp plant debris. The adult flies inflict painful bites that often continue to bleed after the fly has left. The human victim may suffer from fever and general illness. If ever you have seen an ol' cow standing out there, peacefully chewing her cud, then suddenly going berserk with tail flying and head flailing, chances are she has just had a hunk of her neck or rear bitten off by a horsefly or deerfly.

BLOW FLIES, or BLUEBOTTLE and GREENBOTTLE FLIES, are attractive insects; but there the compliments stop. They are thoroughly obnoxious otherwise. They lay their eggs on dead animals, garbage, sewage, or in open wounds of animals. Some related species parasitize and kill animals and even man. Eggs hatch very soon after being laid; larvae are mature in less than two weeks. The short life-cycle mean several generations a season. Their presence inside a residence sometimes indicates that a rodent or bird has died in a crawl space, wall void, chimney flue or attic. Blow flies are strongly attracted to odors of raw or cooked meat, poultry or fish. They are a nuisance indoors as they buzz around the room and bump into windows while trying to get outdoors.

FLESH FLIES are often a problem outdoors. They usually do not come indoors because the adult females prefer to deposit live larvae (no eggs) on meat scraps and dog excrement; therefore, they prefer areas around dog kennels and runs. Flesh flies are similar to houseflies, but they are slightly larger and have characteristic gray and black checkerboard markings on the abdomen or tail section.

Housefly and Face Fly = Order Diptera, Family Muscidae
Fruit Flies = Order Diptera, Family Tephritidae
Horseflies and Deerflies = Order Diptera, Family Tabanidae
BlueBottle Flies = Order Diptera, Family Calliphoridae
Greenbottle Flies = Order Diptera, Family Calliphoridae
Flesh Flies = Order Diptera, Family Sarcophagidae

A combination of sanitation efforts, insecticide spraying, and fly bait use should effectively control flies. The invention of cheap, mass-produced fly screening was one of man's greatest achievements toward assuring relief from flies. Indoor fly control should be 99% exclusion by screening, caulking, etc.

If large numbers of flies are present in your yard, then there is probably a fly breeding site on your property or nearby. Clean up these areas and make whatever changes are necessary to prevent this condition from recurring. If the source of the flies is not on your property, try to locate it and have the responsible persons help solve the problem. If cooperation cannot be obtained from the other person, contact your municipal or state health department to seek their assistance. Do not suffer with a serious, persistent fly problem that results from someone else's neglect.

The first step to control any fly problem successfully is to determine where the fly maggots are breeding. This process, termed source reduction, is always the most efficient method of control. Since flies feed and lay eggs in areas such as garbage cans or dog kennels which are sources of food odors, cleaning these areas eliminates the problem. Control low fly numbers by using a residual insecticide spray to treat surfaces where flies usually rest. Dursban, Cygon or Baygon sprays are all good. Read the label. For fast knockdown, but with no residual control, aerosol fly sprays containing synergized pyrethrins, tetramethrin or resmethrin are worthwhile. These controls are appropriate for fast results and are safe applications around people, pets and food, before or during picnics and outings.

House Fly adult
(IV-23-V VWR)

House Fly — the whole * # ? family
(C-USDA-199)

Common House Fly/Lesser House Fly comparison
(IV-23-U VWR)

Oh, what big eyes you have, Face Fly!
(C-USDA-224)

FRUIT FLY

Vinegar Fruit Fly adult
(IV-27-T VWR)

Vinegar Fruit Fly larvae in fruit
(IV-27-F VWR)

Horse Fly

Horse Fly adult
(DI-9-A VWR)

Black Deer Fly
(C-USDA-214)

Blue Bottle Fly adults
(IV-25-A VWR)

Mediterranean Fruit Fly females on a California lemon. (These Ladies of the Lemon are not from Texas, but they would like to be.)
(DI-1-A VWR)

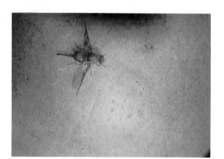

Mexican Fruit Fly female doing her mama duty on Texas citrus
(J. Victor French)

SYRPHID FLIES

I would like you to meet the Syrphid Fly. Now, here's a guy who ain't mad at nobody. He doesn't bite. He doesn't sting. He doesn't chew up leaves or eat anybody. He spends his time hovering over the flowers, sipping nectar. Adults are rather attractive black-and-yellow flies ranging up to ¾ inch in length. They somewhat resemble bees, but with a major difference. A syrphid fly has two wings, a bee has four. And as the syrphid fly indulges in his nectar nipping, he inadvertantly does a super job of pollinating.

Syrphid fly kids, you won't take to: they are slimy, green maggots. Although these things are as ugly as sin, they do a super job of eating aphids. Larvae are usually found on leaf surfaces, often in the midst of a colony of aphids, chomping away. Tolerate them, slime and all. They are your friends. *

Syrphid Flies = Order Diptera, Family Syrphidae

_____ **BENEFICIAL**_____

Syrphid Fly adults on flowers
(Di-10-H)

Syrphid Fly larva feeding on an aphid
(100-1-OO VWR)

TACHINID FLIES

Tachinid Flies are beneficial insects; and because these flies are prolific, their value as parasites is even greater. More than 1,400 species have been described. All species of the family Tachinidae are parasitic on other insects ... what a family of good-guys. Most Tachinid Flies resemble overgrown houseflies; they are grey or brown or black, but without bright colors. Eggs are usually glued to the skin of the host insect, but are sometimes laid on foliage where the insect will eat them along with the leaf. The larvae feed internally on their hosts which almost always die. One species parasitizes over 100 different caterpillar pests. Common prey are hornworms, cutworms and armyworms. Long Live the Tachinid Fly . . . Long Live the Tachinid Fly . . . God Save the Queen! Cheers . . . Cheers . . . Cheers.

Tachinid Flies = Order Diptera, Family Tachinidae

_____ **BENEFICIAL**_____

Tachinid Fly doing a number on a horn worm
(Ento.-TAEX)

_____ Galls (Insect-Induced)_____

Galls are abnormal swellings of plant tissue caused by certain insects, bacteria, fungi, mites or nematodes. Among the insects causing galls are certain moth caterpillars, beetles, flies, psyllids (jumping plant lice), aphids and small wasps. Insect-induced galls are the most common galls in urban areas.

It is important to point out that gall tissue is a plant-product formed in response to a specific stimulus received from an insect, and once formed, these growths do not continue to utilize host plant nutrients; they are not parasitic. Galls are found most commonly on stems and leaves, but also occur on flowers, fruits and on the trunk. Evidence to date suggests that insect-induced galls in Texas do not result in significant plant injury. The exception is found in the occasional damage caused by phylloxera on pecans.

Gall-making Aphids = Order Homoptera, Family Eriosomatidae
Gall-making Flies = Order Diptera, Family Cecidomyiidae
Gall-making Wasps = Order Hymenoptera, Family Cynipidae

CONTROL CLUE

Unless chemicals can be applied when gall-inducing insects fly or when adults are actively depositing eggs, they offer no effective means of control. Systemic insecticides have not proved effective. Once the gall begins its development, it is impossible to stop or reverse its growth with chemicals. In the case of ornamental oaks, the use of sprays should be avoided during late spring and mid-summer, since beneficial parasitic wasps are primarily active during these two periods. In the case of persistent branch and trunk galls, emergence of the adult gall-insect leaves a cavity that is subsequently occupied by beneficial critters. For example, when the adult emerges from the mealy oak gall, the vacated space is taken over by small spiders, lacewing larvae, ants or beneficial wasps. Thus, old galls provide protection for some good guys, which in turn eat the bad guys. In other words, find something else to worry about.

Gall on live oak caused by Cynipid Wasp
(Ento.-TAEX)

Stem gall on cottonwood caused by poplar petiole gall aphid
(Ento.-TAEX)

Galls on pecan caused by Phylloxera
(Ento.-TAEX)

Leaf gall on azalea
(C-USDA-372)

Leaf gall on camelia
(C-USDA-369)

Ocellate maple leaf gall
(C-USDA-158)

Cynipid wasp gall on water oak
(C-USDA-153)

Live oak gallfly caterpillar damage and gall
on coast live oak
(HY-10-A VWR)

Mealy Oak gall wasp asexual female on oak twig
(HYG-18-C VWR)

Mealy Oak gall wasp ovipositing into oak bud
(HYG-18-E VWR)

Grasshoppers

Grasshoppers feed on a wide range of plants including vegetables, but of approximately 600 species occurring in the United States, few are of economic importance. Periodically, when conditions are right, grasshoppers increase to tremendous numbers and devour every green plant in their path. This occurrence is not the rule, but the exception. Early in the Fall of 1983, a horde of grasshoppers invaded downtown Houston. They swarmed over several of the big glass buildings, apparently seeking warmth and shelter. They made a big mess, but were hardly an agricultural threat.

Several grasshopper species will feed on rose bushes. They will devour not only leaves of the plant, but also buds, flowers and stems. Just a few grasshoppers can cause severe damage in a very short time, so act quickly if you see them around cherished plants.

Short-horned Grasshoppers = Order Orthroptera, Family Acrididae
Long-horned Grasshoppers = Order Orthroptera, Family Tettigoniidae

CONTROL CLUE

Since grasshoppers migrate, control might be difficult. If you have a prize rose in bloom, cover it with netting 'til the grasshoppers are under control. If you are forced to use chemical control on your vegetables or roses, use either Malathion or Sevin.

GRASSHOPPER

Short-horned grasshopper adult
(OR-2-A VWR)

Red-legged grasshopper and damage
(C-USDA-81)

Out-of-town grasshopper photographed while attending a convention in Houston
(Ento.-TAEX)

Lubber grasshopper
(C-USDA-129)

Harlequin Bug

The Harlequin Bug is a stink bug dressed up for Saturday night. Adults are red-and-black spotted, are about ⅜ inch long, and are usually mentioned in the same breath with stink bugs because they do the same mischief. These plant bugs will congregate in large numbers on a host plant and literally suck it to death. This Pretty Boy Floyd of the insect world is guilty of everything except bank robbery. He will feed on the fruit of a wide range of plants including beets, squash, beans, peas, tomatoes and corn, causing shriveling and deformity. Don't let the fancy dress mislead you; this guy is a gangster.

Harlequin Bug = Order Hemiptera, Family Pentatomidae

CONTROL CLUE

Remember, if you see harlequin bugs, damage won't be far behind, so hit 'em quick with Sevin spray or the Pyrethrins spray that you used on the other stink bugs.

Harlequin Bug
(C-USDA-89)

Head Lice

AND OTHER UNMENTIONABLES

Note: This page is rated XXX and should not be read if nosy neighbors are over for coffee!

Oh, yes . . . it *can* happen to you. Head lice can happen to anyone. It is not a matter of poor health habits or of being dirty. Head lice are usually transmitted from one infested person to another by direct contact with the hair. Personal items like combs, brushes, towels and bedding are other frequent sources of contamination. Another source of infestation is clothing such as hats, ribbons, scarves, topcoats and sweaters.

Head lice do not observe any class distinction; they can infest anyone and are a problem which is rapidly gaining ground in all strata of our society. The schoolgirl who borrows a comb . . . the student trying on a hat in a fashion shop . . . children who share a bed for a nap . . . the traveller resting his head against the back of an airline seat. That's all it takes. Teachers in our elementary schools are advised to take note of children who immoderately scratch themselves. Head lice are epidemic in our schools. It is a national problem.

Human Head Louse = Order Anoplura, Family Pediculidae

CONTROL CLUE

Although lice are difficult to see, they are easy to recognize provided you're looking for them. Teachers, school nurses and hawk-eyed mothers will readily spot the small, silvery eggs attached to individual hairs. These eggs are called nits. In checking the scalp, pay particular attention to the back of the head and behind the ears. And, mother, if li'l darlin' comes home from school with a note stating he/she has flunked the cootie test, don't crater. It's not the end of the world. Your druggist or your family doctor can tell you what to do. And do it, or your whole family will go down scratching.

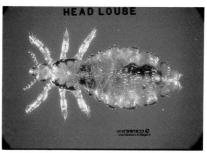
Head Louse
(III-2-B VWR)

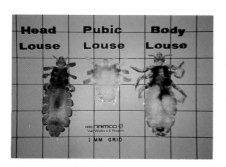
Head Louse/Pubic Louse/Body Louse comparison
(III-4-C VWR)

Head Louse nits on hair
(III-2-G VWR)

Head Louse nits on plastic comb
(III-2-K VWR)

Bedbug
(C-USDA-206)

Bedbug nymph engorged after a blood meal
(III-8-G VWR)

Body louse eggs in seams of clothing —
magnified
(III-3-N VWR)

Body lice and eggs
(III-3-C VWR)

_____ Hickory Shuckworm _____

Hickory Shuckworms frequently cause severe injury to pecan and hickory crops in Texas. Little white worms with brown heads (larvae) infest the shucks covering the nuts. In late summer and fall, the shucks are tunneled out; and as a result of this action, nuts are slower to mature, kernels do not develop properly, and shucks stick to the nuts and fail to open. Larvae overwinter in the fallen pecan or hickory shucks. They pupate in late winter and early spring, emerging as adults during spring and summer.

Adult moths, which are dark brown to grayish-black and are about ⅜ inch long, deposit their eggs mainly on leaves and young nuts. The hatched larvae feed in developing nuts in early summer. Succeeding generations develop in pecan nutlets during June, July and early August and in the shucks during the remainder of the season. As many as five generations may be completed each year before fall larvae go into hibernation.

Hickory Shuckworm = Order Lepidoptera, Family Olethreutidae

Hickory Shuckworm larva
(C-USDA-163)

_____ Hornworms _____

These two critters, the Tomato Hornworm and the Tobacco Hornworm, though genetically different, are so alike in appearance and dirty work, I thought we might save time by considering them as one. First of all, these hornworms come from truly magnificent parents . . . large, fast-flying hawk moths (sometimes mistaken for hummingbirds) that are endowed with a five-inch wingspan. And that's the only nice thing we can say about hornworms.

For the record, the Tobacco Hornworm larvae have seven diagonal light stripes; the Tomato Hornworm larvae have eight curved stripes. Basic colors are green and brown with a few shades in between. A red or black "horn" projects from the rear end. And can they eat! Hornworms feed on the fruit and foliage of tomatoes, peppers and eggplant. A few large larvae can strip a plant before you know it; if allowed to do so, they will grow to the size of a good cheroot cigar.

Incidentally, sometimes these worms may be found with puffed-rice-like white sacs on their bodies. These sacs are the cocoons of parasitic wasps (Braconid Wasp) that feed on and will eventually kill the hornworm. Better you should let the wasp take care of this hornworm and in the process make another batch of parasitic wasps to prey upon more hornworms.

Hornworms = Order Lepidoptera, Family Sphingidae

_____ CONTROL CLUE _____

Your best bet, (other than the wasp): pick 'em off the plant, drop 'em on the ground and step on 'em.

Tobacco Hornworm on tomato
(C-USDA-91)

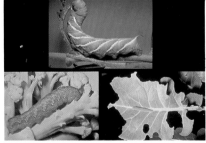

Tomato Hornworm/Tobacco Hornworm/ hornworm damage to leaf
(ORTHO-37)

Lace Bugs

Lace Bugs commonly damage several broadleaved evergreens such as azaleas, laurels, pyracanthas and rhododendrons. Deciduous trees (those that shed their leaves) which may be attacked are sycamore, oak, hawthorn, quince, American elm, black walnut and apple.

Adult lace bugs are ⅛ to ¼ inch long, are flattened and are rectangular in shape. They get their name from the appearance of the area behind the head and from the wing covers. The wings of most lace bugs are transparent. The lace bug nymph is flat and oval-shaped. Both adults and nymphs (babies) have piercing-sucking mouthparts to take sap from the underside of a leaf. Damage appears on the upper leaf surface as white, yellow or brown specks. One of the most common lace bugs in Texas is the hawthorn lace bug, the major pest of pyracantha.

Lace Bugs = Order Hemiptera, Family Tingidae

CONTROL CLUE

If only a few lace bugs are present and little or no damage is observed, wash them off with a strong stream of water from the garden hose. If chemical control is necessary, use Orthene, Sevin or Malathion.

Lace Bug adults
(C-USDA-134)

Lace Bug — magnified
(HE-5-A VWR)

Lace Bug damage to azalea
(C-USDA-135)

_____ Lacewings _____

Here is Miss America of the bug world. In the adult form, she's one of the most beautiful and is beneficial. Adults are up to ¾ inch long, have long antennae and transparent, lacy wings. Lacewings come in several models, generally white, green or brown. Adults don't do a great deal of anything except look pretty, make love, and eat a few aphids here and there; but perhaps their most significant contribution lies in their ugly progeny. The larvae are such voracious eaters that they are commonly called Aphid Lions. They suck the body fluids not only from aphids, but also from mealybugs, scales and other small insects. Mama must lay her eggs singly on top of a delicate hair-like stalk so the bigger kids don't eat 'em. The larvae are rather small (up to a ½ inch), are flat and rather resemble miniature alligators without tails.

If you're blessed with lacewings, you are most likely to find them on your screen door at night if the porch light is on. The adults are attracted to bright lights. Don't spray them with bug stuff. Remember, they are good for you.

Brown Lacewing = Order Neuroptera, Family Hemerobiidae
Green Lacewing = Order Neuroptera, Family Chrysopidae

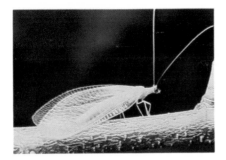

White Lacewing
(C-USDA-185)

Lacewing larva (aphid lion) feeding on an aphid
(100-1-LL VWR)

Green Lacewing adult
(100-1-MM VWR)

Green Lacewing egg on silk stalk
(100-1-KK VWR)

Brown Lacewing adult
(NE-1-B VWR)

Leaffooted Bug

No big deal about Leaffooted Bugs; they are stink bug kinfolks. What's unique about these cowboys is their odd appearance. These large (¾ inch) bugs are generally gray-brown colored with a distinctive white line across the back. Those strange flattened areas on their hind legs might remind you of snow shoes. A leaffooted bug walks like a tomcat stepping through high, wet grass. They will put those yellow zits on your tomatoes and will pig-tail your okra, so don't ignore them.

Leaffooted Bug = Order Hemiptera, Family Coreidae

_____ CONTROL CLUE _____

Dexol Vegetable Garden Spray, which contains Pyrethrins, is what I use. It comes ready-mixed and can be applied on the day of harvest. Read the label.

Leaffooted bug adult
(C-USDA-99)

Leaffooted bug hind leg with those characteristic cowboy "chaps"
(HE-8-F VWR)

Leafhoppers

Several species of Leafhoppers attack vegetable crops in Texas. Some of the more bothersome ones are SOUTHERN GARDEN LEAFHOPPER (known as the BEAN LEAFHOPPER in the Lower Rio Grande Valley); ASTER LEAFHOPPER; POTATO LEAFHOPPER; WESTERN POTATO LEAFHOPPER; and BEET LEAFHOPPER. The Bean and Beet Leafhoppers are green to pale green in color, are somewhat wedge-shaped and are ⅛ to ¼ inch in length. The Aster Leafhopper adult is brownish-gray and has six spots on the face above the antennae. Nymphs are green with wing buds instead of wings, are smaller than adults, but are also the same wedge shape.

Most leafhoppers excrete large quantities of honeydew. Leafhoppers feeding results in leaf stippling, curling, stunting and dwarfing, accompanied by a yellowing, browning or blighting of foliage. Many are vectors of plant diseases. For example, the only damage caused by the Aster Leafhopper is the transmission of the virus disease, Aster Yellows, occasionally evident in South Texas. Leaves of plants that are severely damaged are unsightly and lose much of their ability to produce food.

Leafhoppers = Order Homoptera, Family Cicadellidae

CONTROL CLUE

Malathion or Diazinon formulated for food crops will get 'em. Don't forget to spray the underside of the leaves.

Leafhoppers on leaf
(Ento.-TAEX)

Leafhopper close up
(ORTHO-33)

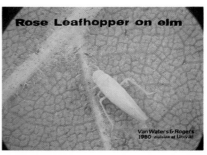

Pine Leafhopper adults (camouflaged)
(HL-8-A VWR)

Rose Leafhopper adult on elm leaf
(HL-9-D VWR)

_____ Leafrollers & Leaf Tyers_____

Leafrollers and Leaf Tyers are small caterpillars that feed inside leaves, which they roll or tie together. They are similar in their habits; but leafrollers roll leaves around themselves, while leaf tyers tie leaves together with silk threads. Larvae vary from pale yellow to dark green; all are about ¾ inch long when mature. Only part of their life may be spent feeding inside the rolled or tied leaves. At other times they may feed inside buds, flowers, or fruit. Pyracanthas, carnations, chrysanthemums, geraniums, roses, zinnias, honeysuckles and verbenas are especially subject to attack, but some species will also attack fruit and hardwood trees.

The unique habit of rolling or tieing those leaves together gives the insects some protection from unfavorable weather, predators and chemical sprays. Leafrollers and leaf tyers may be serious pests in the garden, feeding on many fruits, vegetables and ornamentals. Those species that feed on flowers and fruits are usually much more damaging than those that feed exclusively on leaves.

Leafrollers = Order Lepidoptera, Family Tortricidae
Leaf Tyers = Order Lepidoptera, Family Tortricidae

Leafroller damage on pyracantha
(Ento.-TAEX)

_____ Mealybugs _____

Mealybugs are soft-bodied, sucking insects that are close relatives
of scale insects. Plant parts heavily infested often appear to be covered
with cotton; this appearance is caused by white or gray threads of wax
with which mealybugs cover themselves. They are very active when

young, crawling all over the plant until they find a suitable place to settle, but as the young mealybugs mature, they become sluggish. Mature females move around very little. Adult males do not feed, but die after mating.

Mealybugs may damage any part of a plant by sucking out the sap, which in turn causes leaf distortion, yellowing, stunting, or galls, and ultimately can cause a plant to die. They also coat the plant with large quantities of undigested sap, called honeydew. Most species of mealybugs are garden pests only in warmer climates, but will infest house plants and greenhouse plants in any climate.

Mealybugs = Order Homoptera, Family Eriococcidae

CONTROL CLUE

Mealybugs aren't difficult to control if you can penetrate their waxy-thread cover, so tump some spreader-sticker into your Malathion spray; that should get 'em. If you find only a few mealybugs on a house plant, put a drop of rubbing alcohol on each bug with a cotton swab. This will work nicely if you have the patience to do it.

Golden Mealybug infestation on Norfolk Island Pine close up with eggs, nymphs and adults
(HM-2-B VWR)

Longtailed Mealybug nymphs, adults, wax and black sooty mold close up
(HM-5-H VWR)

Longtailed Mealybug infestation on orange leaf
(HM-5-L VWR)

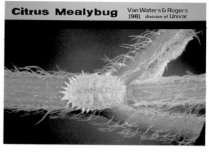

Citrus Mealybug adult magnified
(HM-1-C VWR)

Comstock Mealybug
(C-USDA-124)

Azalea Mealybug
(C-USDA-136)

Mealybug infestation on coleus
(C-USDA-123)

Mealybug destroyer — adult Ladybird Beetle
(101-1-WW VWR)

_____ Minute Pirate Bug _____

Not much need be said about the Minute Pirate Bug. He's a beneficial insect and I don't know anything bad about him. This little guy simply spends his time sucking on insect eggs, aphids and other small soft-bodied pests. He's not very ambitious, is he?

Minute Pirate Bug = Order Hemiptera, Family Anthocoridae

_____ **BENEFICIAL**_____

Minute Pirate Bug adult
(HE-16-A VWR)

_____ Mites (Garden & Landscape)__

Several Mite species attack house plants, garden plants, lawns and trees. In Texas, the most common are Spider Mites and Eriophyid Mites. Showing you pictures of a bunch of different mites would serve little purpose as a prelude to control. Mites are so tiny, you're not likely to ever see them; what you will likely see is the evidence of their presence.

SPIDER MITES are oval shaped and when viewed through a magnifying glass, appear to be sparsely covered with long hairs. Some produce large quantities of fine webbing. The Two-Spotted Spider Mite feeds on a wide variety of plants and is one of the most common mite varieties that attack home gardens and ornamental plants. They usually begin colonies on the undersides of leaves and webbing is visible to the naked eye; all stages of this mite's development occur in and beneath this webbing.

The Carmine Spider Mite is dark red to carmine in color with a dark area on each side. It also feeds on the under surface of leaves. Webbing is usually abundant and high populations may web-over entire leaves.

The Pecan Spider Mite is straw-colored to pale green with several to many dark spots along the back and sides. This species feeds on pecan, chestnut, hickory and oak trees. These mites cause dark brown or liver-colored blotches with moderate to heavy webbing on leaves which will appear dry or scorched.

The Banks Grass Mite attacks grassy plants like garden corn, Johnson-grass and Bermuda-grass lawns. Lawns that are frequently watered are not normally damaged. Early damage to corn is seen as stippling on mature leaves followed by a red coloration at feeding sites. These mites can kill corn at the tassel stage of development.

The Brown Wheat Mite sometimes attacks onions, carrots, lettuce, melons and strawberries. Damage is quite similar to that caused by drought, even where there is ample moisture. Upon close inspection of infested plants, leaves are seen to have a fine mottling; when these plants are observed at a distance, a bronzing or yellowing effect is apparent. This mite is primarily a dry weather pest.

The Clover Mite feeds on a variety of plants including clovers, lawn grasses and ornamental flowers. Populations usually are highest in spring, late summer or early fall. Clover mites produce no webbing. Injury to plants is first perceived as a winding, narrow trail of tiny bleached spots on infested leaves.

The Texas Citrus Mite is tan to brownish green with dark brown spots on the sides. These mites occur year-round on Texas citrus. They appear on upper leaf surfaces, but produce no webbing. Injury to citrus leaves includes stippling, light colored spots and a grayish or silvery overall appearance.

Note: Here's a test for Spider Mites if you don't have a microscope. Hold a clean, white sheet of paper under the sick plant leaf. Briskly thump the leaf several times. You should see several minute specks on the paper. With a pen, draw a tight circle around each speck. Now, wait.

If the specks move out of the circle, then they are alive. Welcome to Spider Mites. Don't sneeze while performing this test. The specks will disappear and you'll have to do everything all over again.

ERIOPHYID MITES, commonly called gall mites, are much smaller than spider mites and produce no webbing. These cigar-shaped critters are difficult to see even with a magnifying glass. They cause many kinds of plant damage. Bermuda-grass Mites infest grass in climates where atmospheric humidity is low. Infested grass is tightly bunched and stunted.

The Tomato Rust Mite attacks many plants including tomato, pepper, potato, eggplant, tobacco and petunia. Damage to tomatoes consists of bronzed stems which may crack longitudinally. Leaves turn brown and may drop if injury is severe; potatoes do not show stem damage, but the leaves dry severely; eggplant can support large populations with damage consisting of distorted or crinkled leaves.

The Citrus Rust Mite feeds on citrus leaf surfaces and on fruit; however, green fruit and the undersides of leaves are preferred.

Spider Mites = Class Arachnida, Order Acari, Family Tetranychidae

---

_____ **CONTROL CLUE** _____

Choice #1 — Kelthane; Choice #2 — Diazinon. Read the label.

Spider Mite damage
(ORTHO-47)

Spider Mite damaged leaves
(Ento.-TAEX)

So you insist upon seeing a Spider Mite.
Okay, okay, already! Here.
(Ento.-TAEX)

Spider Mite damage to holly
(C-USDA-130)

Spider Mites with web and plant injury
(C-USDA-98)

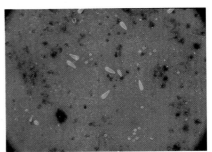

Citrus Rust Mite
(J. Victor French)

Texas Citrus Mite
(J. Victor French)

Two-Spotted Spider Mite damage
(C-USDA-67)

MOSQUITOES

If you're a Texan, I don't have to tell you what a mosquito is. Many different species of this insect pest occur throughout the world. They are as different in their feeding and flying habits as human beings are different in their life styles. Harris County, Texas alone is home for 53 different species, and 35 of these feed on human beings. The remainder feed on large mammals, amphibians, fowl and plants. Only the female bites and she then deposits her eggs in whatever habitat happens to favor her species. All mosquitoes must have water for their lifecycle. In Harris County six types of habitat are enjoyed by the various mosquitoes that are vectors of disease: temporary water (flood and rainpool); salt marsh; permanent fresh water; natural cavities and artificial containers.

SOUTHERN HOUSE MOSQUITOES *(Culex quinquefasciatus)* lay their eggs in artificial containers and in water with high organic content. The female lays her eggs in rafts which may contain 50 to 400 eggs. They are night time biters and rest during daylight hours near their breeding places. They fly only a short distance from their breeding site. This is the mosquito that transmits sleeping sickness (St. Louis Encephalitis).

TIGER MOSQUITOES *(Aedes aegypti)* lay their eggs in artificial containers around human dwellings. They lay their eggs singly on the sides of containers just above the water line to await the next rainfall to provide the water to cover them. The female is a quiet attacker and usually likes to bite around a person's ankles or back of neck. She will bite during the day or in lighted rooms at night. This Aedes is considered the most domesticated and breeds in areas where humans live; it will fly from one block up to 200 yards. This mosquito transmits Yellow and Dengue fever to humans and heartworms to dogs.

TREE HOLE MOSQUITOES *(Aedes triseriatus)* also breed in artificial containers. They are fierce biting mosquitoes, have a short flight range and usually infest wooded areas and dwellings near their breeding containers.

Aedes vexans breed in rainpools, floodwaters and just about any temporary fresh water. They lay their eggs on the ground and hatch after being flooded. Adults may fly 5 to 10 miles a day and are vicious night time biters. The females are attracted to light.

Psorophora columbiae and *Psorophora ciliata* come from rice fields. Both attack day and night. Ciliata is a very large mosquito also known as a galliniper. It is formidable because of its size and vicious bite. Its flight range is about a mile. Columbiae females are smaller than the ciliata, but they can make outdoor activity unbearable for humans and have been known to kill cattle when present in large numbers. Columbiae was considered the vector of Venezuelan Equine Encephalitis in Texas in the year 1971.

Aedes sollicitans and *Aedes taeniorhynchus* lay their eggs on the mud in marshes and other areas which are prone to salt water flooding. Both of these species are strong flyers and have been known to fly 5 or 10 miles and up to 40 miles in search of a blood meal. They are fierce biters both day and night and have been associated with the transmission of dog heartworms.

Anopheles crucians and *Anopheles quadrimaculatus* are Harris County residents also. Crucians breed in the acid type water of cypress swamps, lake margins and slow moving streams. They will bite man, but are not as vicious as other species. Their flight range is more than a mile and they are attracted to light. The quadrimaculatus species vector malaria and breed primarily in permanent fresh water ponds and swamps. They are night flyers with a range of one-half to one mile. They will enter a house to feed on humans, but prefer cattle, horses, mules and chickens. I feel something biting . . . don't you?

Mosquitoes = Order Diptera, Family Culicidae -

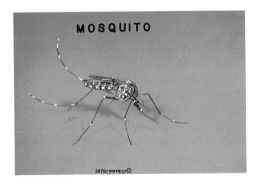

MOSQUITO

Mosquito adult
(III-10-A VWR)

The most effective mosquito control program will be performed by an entity such as your local Mosquito Control District. However, individuals can contribute a great deal by practicing the Ten Commandments of Mosquito Control:

1. Empty, remove, cover or turn upside down any receptacle that can hold water — particularly old bottles and tin cans.
2. Change water and scrub vases that hold flowers or cuttings twice each week. (Or grow your cuttings in damp sand.)
3. Discard old tires or store them indoors.
4. Screen rain barrels and openings to water tanks or cisterns. Seal cisterns not in actual use.
5. Repair leaky plumbing and outside faucets; they may cause standing water.
6. Clean clogged roof gutters and drain flat roofs.
7. Fill holes in trees with sand or mortar, or drain or spray them.
8. Stock ornamental ponds with mosquito-eating fish.
9. Connect open waste-water drains to a sewerage system or construct a separate sump or leach.
10. Fix or replace screens on doors and windows.

Mosquitoes are attracted to people for a variety of reasons — by the carbon dioxide they emit, by the color of clothing, or by perfumes and aftershaves. But the most pervasive reason is for that blood meal necessary for procreation. Dark or blue colors attract mosquitoes, but they seem to stay away from light colors (except blue). Mosquitoes, however, tend to swarm around someone dressed in white at twilight. Yellow light bulbs will have less tendency to attract some species that are attracted to light. Repellents are perhaps the best deterrent for an individual; liquid repellents can protect against mosquito bites for 2 hours or more depending on the person, species of mosquito that attacks, and the number of mosquitoes. Other devices available for mosquito control — electric thermal foggers; engine-driven thermal foggers; electric misters; hand-held portable insecticide foggers; and back-pack insecticide foggers. Remember, in the effort to control mosquitoes . . . we hang together, or we slap alone!

Nantucket Pine Tip Moth

Nantucket Pine Tip Moths occur throughout the pine forests of southern and eastern United States. In Texas, the larvae of this moth are responsible for an immense amount of damage to small pines in plantations, in forests and also in smaller ornamental plantings. Open growth trees less than 15 feet tall are most severely attacked. Of the native Texas trees, short-leaf and loblolly pines are especially susceptible to attack, while slash and longleaf pines are rather resistant to this critter. In most areas of Texas, there are four generations per year with a fifth generation occuring during the most favorable seasons.

Nantucket pine tip moths may damage a tree by causing poor tree shape, by stunting growth, or by reducing cone crops; and, in severe instances, they may even kill the pine. Early evidence of tip moth larvae is difficult to detect and may consist of only an occasional dead needle and small webs. The bud seems to be the choicest morsel. After the bud has been eaten, the larvae will then bore down the center of the young stem.

Nantucket Pine Tip Moth = Order Lepidoptera, Family Olethreutidae

Nantucket Pine Tip Moth adult on loblolly pine needle
(TF-3)

Loblolly pine severely damaged by Nantucket Pine Tip Moth
(TF-9)

Pine shoot dissected to show pupa of the
Nantucket Pine Tip Moth.
(TF-8)

Nantucket Pine Tip Moth larva on pine
shoot. Note two ectoparasites.
(TF-5)

┌─────────── CONTROL CLUE ───────────────────────────┐

Presently, there is no effective control for large forest areas.
Homeowners can reduce tip moth injury by properly watering and fertilizing their ornamental pines and by adequate
applications of a systemic insecticide like Dy-Syston. Spraying with Cygon will do a good job if your timing is right.
Check with your local county extension agent about that.
Also, if heavy rains occur within 2 days after spraying, do it
again.

└──┘

Nematodes

There are some 80,000 varieties of Nematodes. Although some
nematodes are actually beneficial, some 1500 of them are classed
detrimental. A number of these "bad guys" attack many of your ornamental and garden plants as well as lawn grasses. You might identify
one of these by its symptoms, but most likely you will never actually
see this nematode. Root Knot Nematode damage is most noticeable to
the observer, but other nematode types cause similar plant damage with

less obvious symptoms, such as stubby root condition, dead areas within roots, excessively branched roots or death of an entire root branch. And that's what you're up against when you have nematodes!

If your plants are legumes such as beans or peas, don't confuse those nitrogen-fixing nodules with root knot nematodes. Nitrogen nodules, attached to the side of the root, can be removed with the thumbnail without destroying the root, while root knot galls are formed within the root and cannot be removed without root destruction. Nematodes are capable of moving only about one foot per year by their own motion; therefore, they must rely on movement of soil, water or plant material for major distribution. Soil clinging to tillage equipment is one of the most common means of spreading nematodes.

CONTROL CLUE

Don't expect to eradicate nematodes by cultural or chemical means alone; however, a combination of suggested procedures can reduce nematode numbers so that a successful crop may be grown:

1. Plant nematode resistant varieties if available. 2. Till the soil regularly during summer and expose it to heat and drying to reduce populations considerably. 3. Rotate susceptible and less susceptible crops. 4. Don't rely on marigolds or put sugar in the soil. You can do more harm than good. 5. If nematodes get too bad, stop gardening in that spot for a year or two.

I don't feel that available chemical controls are worth the expense and the dangers incurred. Why don't you ask me something easy?

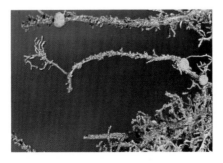

Sting or Lance nematode damage
(C-USDA-240)

Root Knot nematode damage; galled roots.
Lesion nematode damage; sparse roots.
(C-USDA-283)

Nematode damage to corn roots
(C-USDA-249)

___ Oriental Fruit Moth ___

Oriental Fruit Moths are one of two varieties that attack deciduous fruits in Texas. These pests usually are found only in the eastern areas of the state. They attack peaches and plums as well as apples and pears.

The larvae overwinter in cocoons under the tree bark. They feed on buds and succulent new growth in the terminals, creating a condition called "flagging" — the tips of damaged shoots break off. Later in the summer, successive generations attack the fruit directly when no juicy, young foliage is available. Mature larvae leave gum-filled holes in the fruit when they exit to pupate, and several generations can occur in a single year.

Oriental Fruit Moth = Order Lepidoptera, Family Olethreutidae

Oriental Fruit Moth damage to peach twig
(C-USDA-177)

_____ Pantry Pests _____

A variety of insects infest nearly any type of dried food stored in
the pantry. These pests may be brought in as eggs or larvae in pur-
chased food, or adult insects may be attracted to food for egg laying
from nearby infestations within the home or from outdoors.

ANGOUMOIS GRAIN MOTHS are similar in size and color to clothes moths, but the larvae feed only on whole kernels of corn, wheat, popcorn, indian corn decorations, seeds in dry flower arrangements, and bulk stored grains. No woolens. Adult moths will fly around homes in the daytime, while clothes moths shun the light. This is the tiny, fluttering critter that will watch television with you at night and cause your wife to run for moth balls.

GRAIN BEETLES, FLOUR BEETLES, CIGARETTE BEETLES and DRUGSTORE BEETLES are small, reddish-brown to cinnamon-colored beetles that enjoy collecting around light fixtures. The larvae are small, cream-colored worms with dark brown heads. They infest flour, dried pet food, meal, macaroni, cereals, crackers, prepared cake mixes, spices and dried fruits. The adults feed on the same foods. (If you find some of those little brown-headed worms in your flour supply, don't waste the flour. Bake a fast chocolate cake. No one will ever know; I won't tell.)

DERMESTID BEETLES are scavengers on plant and animal products and will feast on leather, furs, skins, dried meat products, woolens and silk materials, cheese, and cereal grain products. Dermestids can be divided into three categories based upon the type of food preferred: LARDER BEETLES and larvae prefer products of animal origin and may infest dried meats and cheese. They are rarely found on foods of plant origin. You will more likely find these critters out in the compost box rather than in the pantry. CARPET BEETLES also prefer products of animal origin, but may be found throughout the home feeding on carpets, clothing, upholstery and wool or silk fabrics. They are occasionally found in stored food products. If you discover carpet beetles in your home, call a qualified pest control service. Carpet beetles can cause you great damage and cost you big money. CABINET BEETLES are the only group of dermestids that prefer cereals, grain products, spices and other true pantry items. The larvae do most of the damage.

FLOUR MOTHS are about ⅓ inch long. Typical examples are the INDIANMEAL MOTH and the MEDITERRANEAN FLOUR MOTH. The larvae will leave their food site and wander around searching for a suitable place to pupate. The moths are strong fliers and often are found flying in proximity to the infestation site.

BROWN SPIDER BEETLES are about ¼ inch long and are equipped with long spindly legs. They rather look like spiders, but they aren't. Both adults and larvae are active feeders.

MEALWORMS, PSOCIDS and GRAIN MITES may be found where foods become moist or moldy. Destroy the infested food and correct the moisture problem.

Angoumois Grain Moth = Order Lepidoptera, Family Gelechiidae
Saw-Toothed Grain Beetle = Order Coleoptera, Family Cucujidae

Confused Flour Beetle = Order Coleoptera, Family Tenebrionidae
Cigarette Beetle = Order Coleoptera, Family Anobiidae
Drugstore Beetle = Order Coleoptera, Family Anobiidae
Dermestid Bettles = Order Coleoptera, Family Dermestidae
Larder Beetle = Order Coleoptera, Family Dermestidae
Carpet Beetle = Order Coleoptera, Family Dermestidae
Indian Mealmoth = Order Lepidoptera, Family Pyralidae
Cabinet Beetles = Order Coleoptera, Family Dermestidae
Mediterranean Flour Moth = Order Lepidoptera, Family Pyralidae
Brown Spider Beetles = Order Coleoptera, Family Ptinidae
Barklice (Psocid) = Order Psocoptera, Family Psocidae
Grain Mites = Class Arachnida, Order Acari, Family Tyroglyphidae
Lesser Mealworm = Order Coleoptera, Family Tenebrionidae

CONTROL CLUE

Simply put, seek out the source and destroy. Thoroughly wash food containers and shelves with hot, soapy water; store fresh food products in your freezer; call your pest control operator for the tough control problems.

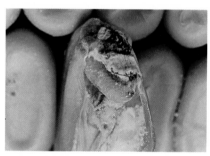

Angoumois Grain Moth larva
(C-USDA-25)

Angoumois Grain Moth adult
(C-USDA-24)

Saw-toothed Grain Beetle
(C-USDA-28)

Flour Beetle adult
(C-USDA-26)

Red Flour Beetle adult
(V-11-M VWR)

Cigarette Beetle
(C-USDA-03)

Dermested Grain Beetle adult
(C-USDA-16)

Dermested Grain Beetle larva
(C-USDA-15)

Cigarette Beetle & Drugstore Beetle adult comparison
(V-8-G VWR)

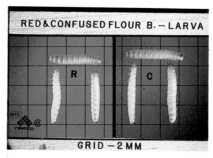

Red Flour Beetle/Confused Flour Beetle larvae comparison
(V-11-B VWR)

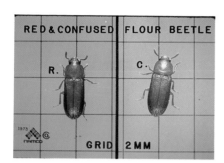

Red Flour Beetle/Confused Flour Beetle adult comparison
(V-11-H VWR)

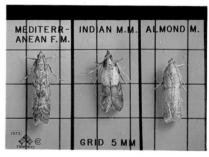

Indian Meal Moth/Mediterranean Flour Moth/Almond Moth adult comparison
(V-1-N VWR)

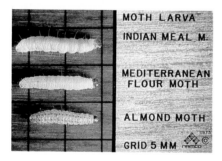

Indian Meal Moth/Mediterranean Flour Moth/Almond Moth larvae comparison
(V-1-G VWR)

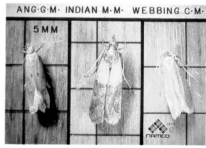

Indian Meal Moth/Angoumois Grain Moth/ Webbing Clothes Moth adults
(V-1-O VWR)

Indian Meal Moth — adult & larva
(C-USDA-13)

Almond Moth larva damage to peanuts
(C-USDA-10)

Brown Spider Beetle larva/pupa/adult
comparison
(V-14-C VWR)

Psocids (book lice)
(V-22-A VWR)

Mealworms — yellow and dark
(C-USDA-20)

Cowpea Weevil adults on blackeyed bean
(V-21-P VWR)

Cowpea Weevil larva inside blackeyed
bean
(V-21-I VWR)

Pecan Catocala

Several species of Catocalas may strip pecan leaves in the spring, leaving only the midribs. When fully grown, the caterpillars are dark gray and are about 3 inches long. They are loopers and become very active when disturbed. Both moths and caterpillars are so well camouflaged they blend with the bark when resting on trees and are frequently overlooked.

Pecan Catocala = Order Lepidoptera, Family Noctuidae

CONTROL CLUE

Although most catocala larvae reach maturity before time to spray for the pecan nut casebearer, most will be killed when that application is made. If per chance they get the jump on you and begin stripping too many leaves, don't wait. Hit 'em with Zolone. If you're lucky, the biggest problem you'll have with this insect is pronouncing its name . . . ka-TOCK-uh-la.

Pecan Catocala larva
(Ento.-TAEX)

Pecan Cigar Casebearer

Pecan Cigar Casebearers may be more or less a problem, varying in incidence and consequent degree of damage from year to year. Mamas are moths. The feeding larvae produce tiny holes in pecan leaves. The most unique feature of this critter is its abode. It constructs a light brown, cigar-shaped case about a ¼ inch long in which it remains throughout development.

The pecan cigar casebearer is not Pecan Public Enemy #1, but it is a most curious little creature that sometimes runs amok and requires specific control measures.

Pecan Cigar Casebearer = Order Lepidoptera, Family Coleophoridae

---

CONTROL CLUE

Insecticides employed in your periodic spray program for control of pecan insects will normally control the pecan cigar casebearer. I use Zolone as the insecticide in my program.

Pecan Cigar Casebearer on twig
(Ento.-TAEX)

Pecan Cigar Casebearer and damage
(Ento.-TAEX)

Pecan Nut Casebearer

The Pecan Nut Casebearer is the major pest of Texas pecans. In early spring, the larvae of the overwintered generation feed first in buds and then in developing shoots, causing both to wither and die. Larvae of succeeding generations feed on nuts during late spring and summer. Severe infestations may destroy your entire pecan crop. Are you paying attention?

Adult casebearers are light gray moths that are about a ⅓ of an inch long. They fly at night and hide during the daytime. Young larvae (worms) are first white to pink in color, but later become olive gray to green and grow to about ½ inch in length. First generation larvae hatch from eggs in 4 or 5 days and migrate to buds below the nuts to feed. After a day or two they enter the nuts and in feeding frequently destroy several or all of the small pecans in a cluster. And if you don't do something early, you can kiss your pecan crop good-bye.

Pecan Nut Casebearer = Order Lepidoptera, Family Pyralidae

CONTROL CLUE

Begin spraying very soon after pollination, when the tiny rice-size nutlets turn brown at the tip. And you must spray more than once. At this stage of growth, I spray Zolone every 15 days and get good control.

Pecan Nut Casebearer damage in terminal growth on pecan tree   (Ento.-TAEX)

Pecan Nut Casebearer adult, pupa and eggs
(Ento.-TAEX)

Pecan nuts damaged by casebearer larva
(Ento.-TAEX)

Pecan Phylloxeras

Pecan Phylloxera produces galls on new pecan growth. Leaves, twigs and nuts may be affected. Phylloxera passes the winter as eggs nestled in bark crevices. By spring the tiny nymphs (babies) emerge and feed on tender young growth, secreting a substance that stimulates plant tissue to develop into galls. Adults are soft-bodied critters closely related to aphids. Upon maturing, adults deposit numerous eggs inside these galls which split after 1 to 3 weeks, thereby liberating the next generation of nymphs. Several generations follow during summer and fall, as long as there is fresh, young growth on the tree.

Incidentally, galls are conspicuous swellings which may be 1/10 to 1 inch in diameter. I understand some curious folks, presuming galls to be a fruit of the tree, have actually nibbled a few. (They taste lousy.)

Pecan Phylloxeras = Order Hemiptera, Family Phylloxeridae

Dormant oil spray recommended for obscure scale will normally handle phylloxera. Apply thoroughly to tree trunks and limbs during the winter when trees are dormant.

Pecan Leaf Phylloxera adult
(HX-1-F VWR)

Galls on pecan leaves caused by Pecan Leaf Phylloxera
(Ento.-TAEX)

Pickleworm & Melonworm

Guess what Pickleworms like best to eat? You lose! Summer squash is tops on their menu, but cucumber and muskmelon are big favorites also. The larvae do the damage by eating on blossoms and vines and by invading the underside of fruits. Larvae (worms), bright green and dotted, grow to about ¾ inch in length. Mama moths have dark brown wing margins which merge into lighter areas toward the center. The abdomen tip is tufted with hairs. The activity of the pickleworm is continuous in the Lower Rio Grande Valley; there may be five generations per year.

Melonworms have much in common with pickleworms. They eat at the same table, feasting on muskmelon, cucumbers and squash. Larvae are also bright green, but have dorsal white stripes running the length of the body; and they are larger, growing to an inch and a quarter in length. They feed on foliage rather than blossoms before tunneling into stems and fruit. Melonworm mamas (moths) have velvety-black wing margins with lighter, pearly-white areas.

Pickleworm and Melonworm = Order Lepidoptera, Family Pyralidae

---

CONTROL CLUE

Liquid Sevin will do it to 'em. Begin treating at first sign of damage and repeat weekly till you clean them out.

Young larva of the pickleworm
(C-USDA-114)

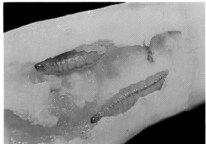

Older pickleworm larva and damage
(C-USDA-115)

_____ Pillbugs & Sowbugs _____

Pillbugs and Sowbugs normally live outdoors. They really are not insects, but are more closely related to shrimp, crabs and crayfish. Because they have difficulty in maintaining body moisture which is necessary for their survival, they habitually remain beneath objects on damp ground or even below ground during the day.

Both of these critters feed on decaying organic matter, but occasionally will feed on young plants or their roots. They are most attracted to over-ripe fruit lying on the ground. In my garden they give my strawberries fits. The compost pile is Thanksgiving dinner every day for these crustaceans; a citrus peeling or a rotten apple can be a gourmet treat.

Pillbugs and sowbugs maintain similar lifestyles, but with one notable difference. When disturbed, pillbugs roll up into a ball; sowbugs do not. They sometimes enter houses, but do no damage.

Pillbugs (sowbugs) = Class Crustacea, Order Isopoda

CONTROL CLUE

A pillbug and sowbug bait containing Carbaryl, Diazinon, Malathion or Methoxychlor is your best bet outdoors. Have no delusions about eliminating these dudes, for you're not likely to do that. Holding the population level down is really the most you can hope for. If they come in your house, pick 'em up with the vacuum cleaner. Residual insecticides are usually not effective in the house.

Pillbug and Sowbug adult comparison
(IV-29-C VWR)

Praying Mantids

Here's a dapper fellow. When I was a kid back in Brazos County, we called this critter a Grandfather's Walking Stick. Wrong name. Now, there *is* an insect called the walking stick; it is a large, usually wingless insect with legs all about the same length. Walking sticks live on and feed on leaves of certain trees, occasionally causing damage. Praying mantids also have elongate, stick-like bodies. Those front legs fit together in order to hold prey.

Mantids range in size from a half inch to over six inches and will feed on just about anything they can hold with their two front legs. They do no damage except to other insects, and that's why they are considered beneficial. They kill their prey by biting the back of the neck, which severs nerves and leaves the victim helpless. They have no qualms about feeding on one another, either. The larger feast on the smaller, and after mating, the female may devour the male. So, if you believe in reincarnation and intend to come back as a praying mantid, be sure you are big and are a female.

Many gardeners actually "sow" mantid eggs to encourage the presence of this critter, but mantids are relatively ineffective as pest controllers because they really aren't heavy feeders. The worst that can be said about the praying mantid is he is just as likely to prey on a harmless insect — or even another beneficial insect, such as a honey bee — as he is on a pest. So, who bats a thousand?

The mantid hunts by waiting or by very slow stalking. When something appetizing passes too close, ZAP, and it's supper. No trouble in identifying the mantid. Nothing else looks like him. Body color usually is green, but some are brown or even pink. He can even turn his head to look around. Don't laugh . . . mantids need love, too.

Praying Mantids = Order Orthroptera, Family Mantidae

_____ **BENEFICIAL**_____

Praying Mantid adult
(C-USDA-192)

Praying Mantids mating
(Ento.-TAEX)

Psyllids

Psyllids are kin to aphids. The winged adults are small (1/16 inch) green or brownish critters that spring into flight with their large hind legs, prompting many to refer to them as "jumping plant lice." Psyllids damage plants by sucking plant juice and by transmitting disease; some cause blister-like galls to form on leaves. The wingless immature psyllids are often covered with white, waxy threads.

Psyllids feed on shoots and leaves of plants, causing distortion, stunting and often tip die-back. They may also cover a plant with honeydew, the undigested plant sap that is excreted as they feed. When infestations are heavy, thousands of adults may swarm around and may even invade your home. They really won't hurt you, but will just "bug" you to death.

Psyllids = Order Homoptera, Family Psyllidae

CONTROL CLUE

Spray fruits and vegetables with an insecticide containing Malathion. Remember, read the label.

Potato Psyllid adult
(HP-5-B VWR)

Common Willow Psyllid
(HP-2-A VWR)

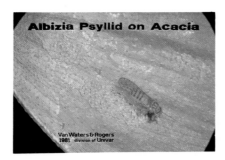

Albizia Psyllid laying eggs on acacia
(HP-1-C VWR)

**Potato Psyllid damage to tomato leaf;
note leaf cupping**
(HP-5-A VWR)

_____ Rodents _____

The ability of Rats and Mice to survive in mankind's world is legend. They exist world-wide and will readily invade well-maintained suburban residences as well as garbage dumps. Evidence of a rat or mouse in your home should be cause for immediate action. The species that most frequently infest houses are the Norway Rat (also known as the house, wharf, brown or sewer rat) and the Roof Rat.

Young rats can squeeze through an opening as small as ½ inch wide. Mice will enter opening slightly larger than ¼ inch wide. Rats will breed in any secluded location outdoors, including heavy vegetation such as ivy or juniper groundcovers. They may burrow in the ground or climb your pecan tree and eat their fill every night. They will invade your garden and munch on tomatoes or eggplant and have strawberries for dessert. If you tolerate their presence outdoors, it will be only a matter of time till you find them feasting on leftovers on your kitchen table.

If allowed to proliferate, rats and mice can pose a serious health hazard as well as be a significant economic problem, in that they both contaminate food with fecal droppings and urine and destroy much property by gnawing. The long front teeth of rats grow constantly and to keep them worn down, rats will gnaw on most anything, including clothing, wood, electrical wires or furniture. Rats occasionally bite people, especially sleeping infants. A rodent bite is serious and should be treated by a physician. Although we normally don't think of squirrels as being rodents, they are and they will do mischief similar to that of rats. Squirrels also will invade your home, but will more likely set up shop in the attic rather than the kitchen.

Rodents = Phylum Chordata

_____ **CONTROL CLUE** _____

Rat or mouse inside the house . . . try a trap and bait it with bacon or peanut butter. Rats or mice in out-buildings . . . offer dry cereal anticoagulant baits in ¼ or ½ pound packets. For small buildings with just a few rats, 2 pounds should be sufficient. For many rats, use 4 or 5 pounds. Use at least 3 pounds for a residence. The average amount for a Texas farmstead is about 10 pounds. Mouseproof your house by sealing the cracks. Control of rodents is really not this simple. Why don't you call a qualified pest control service. Squirrels . . . a pellet gun and a keen eye, but shooting might be illegal in your community, so take care. (I have a great Squirrel Stew recipe. I got it from Grandma Zak.)

Norway Rat
(VR-1-A VWR)

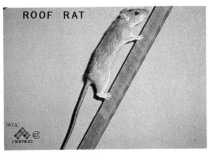

Roof Rat
(VR-2-A VWR)

House Mouse in stored food
(VR-3-A VWR)

Guess who came to dinner? Roof rats!
(VR-2-O VWR)

Tree Squirrel on ground
(VIA-3-C VWR)

Ground Squirrels
(VIA-2-O VWR)

Root Maggots

Not much we can say about Root Maggots except that we have them in Texas and they are a serious problem. Maggots are the larvae of certain flies, and there are several species that prefer certain of our garden crops. There's the CABBAGE MAGGOT that primarily attacks cabbage, cauliflower, broccoli, radishes, beets and other similar crops. The ONION MAGGOT specializes in onions. The SEEDCORN MAGGOT will attack a variety of vegetable crops including corn, beans, peas, melons, cabbage, potatoes and turnips. Also included in the maggot passing parade — CARROT RUST FLY and SUGAR-BEET RUST FLY.

Maggots are usually white, sometimes bearing a yellowish cast. Larvae are about ¼ inch long when full grown. They work in the soil, chew off small roots, and bore into the roots and underground stems of larger plants. Enough about maggots — they're no fun anyway.

Cabbage Maggot = Order Diptera, Family Anthomomyiidae
Sugarbeet Root Maggot = Order Diptera, Family Otitidae

CONTROL CLUE

Work some Diazinon granules into the soil and hope for the best.

Maggots on cabbage root
(C-USDA-104)

Rose Insects

The "superstar" of the flower garden is still the Rose. Many species of insects also find roses irresistible. Sucking insects insert their mouthparts into plant tissue and suck out the juices; chewing insects actually chew on plant tissue, thereby damaging all parts of the rose bush including roots, stems, leaves, buds and open blooms. Not all insects that frequent roses are damaging. Many are incidental; some are pollinators; others are beneficial because they actually attack and feed on rose enemies. To control the enemies, rose growers must first identify them. Perhaps this page will be helpful.

Symptoms of sucking pests:
1. Wilting
2. Presence of honeydew
3. Curling and stunting of leaves
4. Yellowing of foliage
5. Dead spots in tissue

Symptoms of chewing pests:
1. Wilting (root damage)
2. Girdling of stem or cane
3. Severed stems, leaves or buds
4. Holes in leaves or stems
5. Discolored leaves or petals

ROSE APHIDS and other species relish roses. Species vary in color and may be black, green, yellow or pinkish. Watch for black, sooty mold on leaves. Plants appear wilted, and sometimes leaves will yellow and drop or curl and be stunted. Buds may fail to open, be deformed, or produce small blossoms.

LEAFHOPPER adults and nymphs cause a stippling of the leaves. Adults vary in color from gray to yellow and green; some species have patterned markings. Nymphs resemble adults, but are lighter and are wingless. In Texas, leafhoppers may be found on roses from early spring until late fall.

Several SCALE INSECTS attack roses, but the most damaging is the soft Rose Scale. Female rose scales are round and dirty-white; males are elongate and snow-white. Mature scales insert their mouthparts into the plant tissue and remain fixed. They are most abundant with high humidity and reduced sunlight. They not only spoil the plant's appearance, but greatly reduce plant vigor.

Adult WHITEFLIES look like fluttering dandruff when disturbed. Immature whiteflies attach themselves to the underside of leaves and look much like scale insects. Both feed on roses and cause yellow spots on leaves. Heavy infestations can cause defoliation. Watch for that black, sooty mold.

TWO-SPOTTED SPIDER MITES are the most common of several species that attack roses. Their feeding-punctures appear as tiny light-colored spots, giving leaves a stippled appearance. Leaves of heavily infested plants turn yellow, then brown and eventually fall off.

FLOWER THRIPS, ONION THRIPS and TOBACCO THRIPS are the most common species that attack roses. Some are yellow to golden in color; others are almost black. Damaged tissue develops a silvery appearance; heavy infestations result in discoloration and deformed growth. Heavy feeding prevents buds from opening properly and results in deformed, blemished petals.

Leaf-feeding beetles are a continuous problem for rose growers. Watch for: ROSE CHAFERS resemble June beetles and are brown in color. They can destroy a rose bloom in short order. ROSE LEAF BEETLES are small metallic-green critters that feed in buds and on flowers. They riddle the blossoms with holes. TWELVE-SPOTTED CUCUMBER BEETLES feed on tender rose leaves and buds.

LEAF and FLOWER FEEDING LARVAE are incidental pests, but most are voracious feeders. Only one or two can cause extensive damage. Good examples are the Corn Earworm and the Black Woolly Bear.

ROSESLUGS are the immature stages of sawflies. Small larvae usually skeletonize leaves; larger larvae likely will eat the whole thing.

LEAFROLLERS are small, pale-green larvae with black heads that feed inside leaves that they have rolled up and tied with silk threads.

LEAF TYERS are similar to leafrollers in that they draw several leaves or parts of leaves together and tie them with silk. They are light-green in color and are about ¾ inch long when full grown. They eat leaves.

ROSE STEM BORERS attack the stems or canes of roses. They are white larvae that bore into the sapwood and often will girdle the canes in several places. Some infested canes die back to the girdled area; others develop a swollen or enlarged area at the point of injury. Infested stems should be cut and destroyed.

LEAFCUTTING BEES, the phantoms of rose-land, are solitary bees that cut those circular pieces from rose leaves. The nature of damage is obvious.

GALL WASPS lay their eggs in rose stems. When the eggs hatch and young larvae begin to feed, the plant forms an enlarged area, or gall, in the damaged area. Infested tissue should be removed and destroyed before the wasps complete their development.

ROSE MIDGE is a serious problem for rose growers. The adult is a small yellowish fly. Female flies lay eggs in the growing tips of rose stems. The young larvae, or maggots, feed on the tender tissue, kill the tips and deform the buds. Remove and destroy infested tips daily to prevent maggot development.

GRASSHOPPERS feed not only on the leaves, but also on the buds, stems and blooms of roses. One or two grasshoppers can give your rose bush a rough time. Since grasshoppers migrate, control will be difficult.

Rose Aphid = Order Homoptera, Family Aphididae
Leafhoppers = Order Homoptera, Family Cicadellidae
Softrose Scales = Order Homoptera, Family Coccidae
Whiteflies = Order Homoptera, Family Aleyrodidae
Two-spotted Spider Mite = Class Arachnida, Order Acari, Family Tetranychidae
Thrips = Order Thysanoptera, Family Thripidae
Rose Chafer = Order Coleoptera, Family Scarabaeidae
Rose Leaf Beetle = Order Coleoptera, Family Chrysomelidae
Spotted Cucumber Beetle = Order Coleoptera, Family Chrysomelidae
Corn Earworm = Order Lepidoptera, Family Noctuidae
Roseslug = Order Hymenoptera, Family Tenthredinidae
Leafrollers = Order Lepidoptera, Family Tortricidae
Leaf Tyers = Order Lepidoptera, Family Tortricidae
Rose Stem Borers = Order Coleoptera, Family Buprestidae
Leafcutting Bees = Order Hymenoptera, Family Megachilidae
Gall Wasps = Order Hymenoptera, Family Cynipidae
Rose Midge = Order Diptera, Family Cecidomyiidae

Rose Curculio adult and extensive damage on rose bud close up
(CW-7-D VWR)

Rose Aphid winged migrant & apterous female with pink nymphs close up
(HA-29-G VWR)

Rose Aphid infestation on rose bud
(HA-29-B VWR)

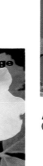

Rose Curculio adults & damage to rose petals of rose in bloom
(CW-7-G VWR)

Western Rose Chafer adult on rose
(CJ-4-B VWR)

Leafcutting Bee damage to rose leaves
(HY-18-B VWR)

A Rose Chafer
(C-USDA-182)

Fuller's Rose Weevil and damage to eucalyptus leaf
(CW-4-B VWR)

Black Woolly Bear, just one of the wandering critters that will strip a rose bush in short order
(C-USDA-141)

Bristly Rose Slug mature larva, dorsal &
coiled view on rose leaf
HYS-3-C VWR)

Bristly Rose Slug extensive damage to rose
bush
(HYS-3-H VWR)

_____ Sawflies _____

Sawflies are a large and diverse group of insects that defoliate hardwood and coniferous trees. Several species are native to Texas, but the ones that cause the most concern are the RED-HEADED PINE SAWFLY and the BLACK-HEADED PINE SAWFLY. They are both important defoliators of young southern yellow pines.

Sawflies are not true flies; rather, they are wasps that cannot sting. They get their name from the saw-like structure on the end of the female's abdomen which she uses to lay her eggs. It is the larvae that consume the needles or leaves and cause damage. They usually feed on a tree from top to bottom, completely defoliating one branch before moving to another. One variety feeds on pecan foliage during April and early May. Occasionally, sawflies are a problem to ornamental trees.

Fertilized eggs produce both male and female offspring, but the unfertilized eggs produce only females. (Try to figure that one out).

When weather conditions are favorable, as many as five generations may be produced in one year. Full grown red-headed pine sawfly larvae are about an inch long. They have a red-rust head and a hairless, yellowish-white body with six rows of black dots. Black-headed pine sawfly larvae are marked differently, but the most distinguishing difference is the black head.

Sawflies = Order Hymenoptera, Family Tenthredinidae

CONTROL CLUE

Sprays should be applied to the foliage as soon as an infestation is observed. Methoxychlor, or Sevin, or Zolone can do the job, but you might need to spray more than once. Read the label.

Black Headed Pine Sawfly larvae
(PD-14)

Red Headed Pine Sawfly larvae
(C-USDA-144)

Oak Sawfly larva
(C-USDA-154)

Scale Insects

Homeowners may have difficulty in controlling scales because they tend to lose sight of the fact that scales are insects. Mature scales *never move* once they firmly attach themselves to branches, twigs or foliage.

There are hundreds of scale species found on a variety of host plants. A general differentiation is SOFT SCALES and ARMORED SCALES. Oval, soft-bodied Mealybugs, while not true scales, are closely related and will respond to scale controls.

Scales are weird lookin' critters which easily go unnoticed until both the infestation and consequent damage are extensive. At this point, the uninformed homeowner is likely to push the panic button and may go after this insect with a fungicide, or may apply the proper insecticide at the wrong time, or may apply the wrong insecticide at the right time. So, pay attention. Scales weaken or kill the host plant by sucking plant sap through piercing-sucking mouthparts. If scale populations increase, treat plants with insecticide oils during the dormant season, or treat with conventional sprays in the spring and summer. If you don't know what to do at a given time, ask a qualified person for advice.

Soft Scales = Order Homoptera, Family Diaspididae
Armored Scales = Order Homoptera, Family Diaspididae
Mealybugs = Order Homoptera, Family Pseudococcidae

CONTROL CLUE

Use a dormant oil, or a conventional insecticide (Diazinon), or a contact insecticide (Pyrethrins), or a systemic insecticide (Di-Syston). The nature of the host plant will determine which control should be used. Remember, read the label and don't be afraid to ask. Successful control can be determined by sliding your thumbnail across a group of scales. If they are dry, hollow and flake off readily, they are dead. Some scales can be effectively controlled by daubing with rubbing alcohol or a mild soap solution.

Black Scale crawlers on oleander
(HS-3-C VWR)

Black Scale (nymphs & adults) on oleander branch
(HS-3-D VWR)

Barnacle Scale female and nymphs
(HS-1-B VWR)

Brown Apricot Scale on branch
(101-1-S VWR)

Cactus Scale
(101-1-KK VWR)

Euonymus Scale on euonymus stem
(C-USDA-122)

Cottony Cushion Scale
(C-USDA-120)

Cottony Cushion Scale — ventral view
(101-1-GG VWR)

Cottony Camellia Scale on holly plant
(HS-11-A VWR)

Cottony Camellia Scale on holly leaf close up
(HS-11-B VWR)

Cottony Maple Scale on maple
(HS-13-A VWR)

Ant collecting honeydew from immature Ice Plant Scale
(101-1-U VWR)

California Red Scale on orange branch close up
(HS-10-F VWR)

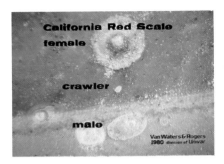

California Red Scale on orange leaf close up — crawler, male/female
(HS-10-H VWR)

Typical infestation of Euonymus Scale on a euonymus plant
(Ento.-TAEX)

Oystershell Scale on branch
(HS-34-A VWR)

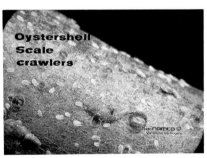

Oystershell Scale Crawlers
(HS-34-D VWR)

Oystershell Scale on Black Walnut
(HS-34-B VWR)

Oak Pit Scales magnified
(HS-33-C VWR)

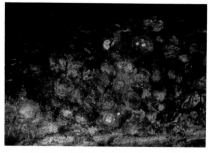

Obscure Scale on Pecan close up
(Ento.-TAEX)

Obscure Scale on Pecan limb
(Ento.-TAEX)

Ivy or Oleander Scale & Soft Brown Scale
on ivy
(HS-43-A VWR)

Peony Scale on azalea
(C-USDA-125)

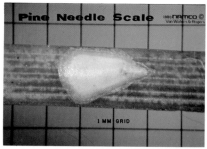

Pine Needle Scale magnified
(HS-36-D)

Pine Scale
(SI-13)

Pine Needle Scale
(SI-19)

Tuliptree Scale
(SI-18)

Pine Leaf Scale on cedar, male and female
(HS-37-E VWR)

Soft Brown Scale on wisteria
(HS-43-B VWR)

Cottony Maple Scale
(SI-16)

San Jose Scale magnified
(HS-42-C VWR)

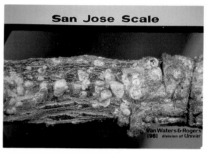

San Jose Scale close up
(HS-42-B VWR)

California Red Scale and immature Soft Brown Scale on orange magnified
(HS-10-D VWR)

Tea Scale on camellia: white — males; dark — females (Under-leaf surface)
(C-USDA-138)

Tea Scales on camellia, upper leaf
(C-USDA-137)

Wax Scale on branch
(C-USDA-139)

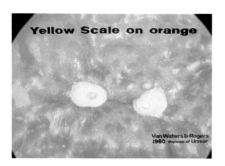

Yellow Scale on orange magnified
(HS-47-B VWR)

Cottony Camellia Scale egg sac opened to
show egg chambers
(HS-11-D VWR)

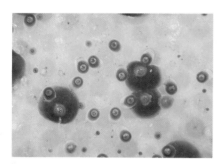

Black Scale nymphs and expired female
scales
(101-1-DD VWR)

Florida Red Scale and damage on citrus
close up
(J. Victor French)

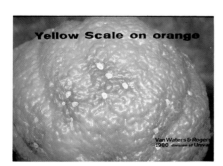

Lac Insects on mesquite, close up
(HS-29-B VWR)

Yellow Scale on orange fruit
(HS-47-A VWR)

Scorpions

Scorpions occur throughout Texas and should be easily recognized by their large pinchers near the head and by their thin tail carried over the back. They range in size from 1 to 5 inches, depending on the species, with colors ranging from yellowish-brown to black. The arched tail ends in a bulb-like poison gland equipped with a stinger. And that's what can hurt you. Forget those pinchers. They are used only for holding food, which might be small insects, spiders, centipedes, other scorpions or earthworms. Scorpions hide under stones, bark of fallen trees, boards, firewood or other objects that lay on the ground.

Although individual reactions to the stings may vary, it is important to seek medical assistance immediately if a person, particularly a child, has severe reaction to a scorpion sting. Ice packs or alcohol swabs applied to the sting area are normally the suggested first-aid treatments.

Scorpions = Class Arachnida, Order Scorpionida

CONTROL CLUE

Chemical control for scorpions is not particularly effective. If you encounter one outdoors, hit it with a rock. If you see one indoors, step on it, but for god's sake, be sure you're wearing shoes!

Scorpion   (C-USDA-212)

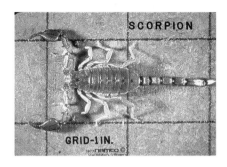
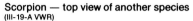

Scorpion — top view of another species
(III-19-A VWR)

Devil Scorpion in hunting posture
(III-19-T VWR)

_____Serpentine Leafminers_____

Leafminers are immature forms of a small fly. The babies (maggots) feed between the upper and lower leaf surfaces, tunnel out the tissue and leave slender, white winding trails through the leaf's interior. You seldom notice the adults that cause the problem. They are small flies which are about an eighth of an inch long, have a yellow and black thorax and a black head. Ten to twenty generations occur per year down in the Lower Rio Grande Valley. Fewer generations per year occur in northern Texas where the growing season is shorter.

Leafminers have the hots for peppers, all cucurbit crops, beans, southern peas, tomatoes, potatoes, spinach, eggplant and other plants. You'll likely not want to eat the damaged portions of edible leaves, but yields are usually not affected on fruit-producing vegetables unless many of the leaves are damaged.

Serpentine Leafminers = Order Diptera, Family Agromyzidae

An ounce of prevention is about the only cure. Once this critter begins doodling on your tomato leaves, it's a little late to start a cure. About all you can do is get an early start with a Diazinon spray and keep your fingers crossed.

Serpentine Leafminer damage to tomato leaf
(C-USDA-108)

Sharpshooters

These are the clowns that play hide-and-seek with you. Sharpshooters are large, active leafhoppers that will avoid humans by running around a plant stem. They can move like a cursor on a computer screen, running backwards and sideways as rapidly as forward. As leafhoppers go, these are king size, growing up to ½ inch in length. They are winged and are very capable flyers. They suck plant juices from a wide variety of plants, but are seldom abundant enough to justify control.

Leafhoppers = Order Homoptera, Family Cicadellidae

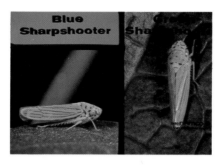

Blue & Green Sharpshooter leafhoppers
(HL-1-A VWR)

_____ Silverfish & Firebrats _____

Silverfish and Firebrats are common house-invading pests. They are fast runners and are most active at night. They can go for long periods of time, sometimes over a year, without food, but will readily feast on flour, dried meat, rolled oats, paper and even glue. Considered mostly a nuisance, they damage paper goods, stain cloth and contaminate food.

Adults reach a length of about ¾ inch. Silverfish are covered with fine scales which are silvery to brown in color; firebrats are quite similar to silverfish but are considerably darker in color. They are usually brought into the home in foodstuffs, furniture, old books and papers.

Silverfish and Firebrat = Order Thysanura, Family Lepismatidae

---

_____ CONTROL CLUE _____

These critters prefer the dark, so give them light. Spraying with Dursban, Diazinon, Pyrethrins or Malathion should handle minor occurences, but when infestations are large, persistent and hard to find, call your pest control operator. He has the knowledge, training and equipment to perform safe and effective control. Take my word for it.

Silverfish
(C-USDA-202)

Typical silverfish damage
(C-USDA-18)

Firebrat

VanWaters & Rogers
1981 division of Univar

Firebrat adult, close up
(VF-8-I VWR)

Snails & Slugs

Snails and Slugs are not insects — these critters are mollusks. They're related to clams, oysters and other shellfish, and like the rest of the family, they must be moist all the time; consequently, they avoid direct sun and dry places. During the day, they hide under anything lying on the ground or they may secret themselves in ground-cover, weedy areas, or compost piles. They emerge at night or on cloudy days to feed.

Snails and slugs are similar except that the snail has a hard shell into which it withdraws whenever it's pleased to do so. Snails and slugs often damage vegetables and garden crops. They feed on tender vegetation such as lettuce or Bird's-Nest Fern. Stems and leaves may be sheared off and eaten. Silvery trails winding around on plants and nearby soil are a definite clue that snails or slugs are present. Inspect the garden for them at night by flashlight.

Snails = Phylum Mullusca, Class Gastropoda
Slugs = Phylum Mullusca, Class Gastropoda, Order Pulmonata

Black Slug
(IV-34-A VWR)

Brown Garden Snails
(IV-33-A VWR)

_____ Southern Corn Rootworm _____

Here's a tricky one. Mama is the Spotted Cucumber Beetle and is
known to feed on over 200 different plant species including many
cultivated crops. In Texas there may be three or more generations in a
year, and this critter is everywhere.

Let me tell you about Mama. She's about ¼ inch long, greenish
yellow in color and has 12 black spots on her back. The adults also have
a prominent head with relatively dark antennae. Adults are general
foliage feeders and often occur in large enough numbers to cause considerable leaf damage. Damage appears as irregular holes in leaves,
damaged growing tips and occasionally the girdling of seedlings at or
near ground level.

While the Spotted Cucumber Beetle takes the high road, the baby
(Southern Corn Rootworm) takes the low road. They feed on plant roots
and bore into germinating seed, large roots and underground stems.
These slender larvae are white to yellowish in color and are about ½
inch long when full grown.

Although genetically different, very close kinfolks are the Northern Corn Rootworm and the Western Corn Rootworm. For your purpose, when you've seen one . . . you've seen 'em all. (Unless you're a Corn Rootworm.)

Southern Corn Rootworm = Order Coleoptera, Family Chrysomelidae

```
_____ CONTROL CLUE _____

Sevin or Diazinon for the old folks (adults) . . . Diazinon
granules for the worms.
```

Southern Corn Rootworm larva
(C-USDA-30)

**Twelve-spotted Cucumber Beetle, adult of
Southern Corn Rootworm**
(C-USDA-31)

Twelve-spotted Cucumber Beetle
(CC-5-B VWR)

Spiders

Spiders enjoy about as much acceptance in society as did the Hunchback of Notre Dame. With only a few exceptions, these critters are not only harmless, but are beneficial. Although all spiders are capable of injecting venom when they bite, only a few, such as the Brown Recluse and the more infamous Black Widow are dangerous to people.

All spiders are beneficial predators on smaller creatures, and many are quite effective at reducing pests. They all employ some sort of webbing to do their thing, and people also tend to resent that web. Brother spider just can't win, can he!

TARANTULAS, JUMPING SPIDERS and WOLF SPIDERS are frequent victims of this spider-phobia and are considered to be dangerous simply because they are ugly. They are large, hairy and formidable, but their bite is less harmful than a bee sting. Some people, however, are extremely allergic to spider venom, so if you are bitten and have an adverse reaction, don't hesitate to see your doctor immediately.

I really don't think you will be able to do this, but the next time you encounter one of these beneficial spiders . . . give the guy a break. Leave him alone.

Tarantulas = Class Arachnida, Order Araneida, Family Theraphosidae
Jumping Spiders = Class Arachnida, Order Araneida, Family Lycosidae
Wolf Spider = Class Arachnida, Order Araneida, Family Salticidae

BENEFICIAL

Clubionid Hunting Spider from Dallas, Texas
(AR-9-A VWR)

Argiope Garden Spider from Houston, Texas
(AR-11-C VWR)

Yellow Crab Spider on a leaf
(AR-8-A VWR)

White Crab Spider
(Ento.-TAEX)

Orb Weaver Spider
(AR-11-A VWR)

Daddy-long-legs
(100-1-TT VWR)

Trapdoor Spider trap-door in soil with
Trapdoor Spider competely out of its
burrow
(AR-4-G VWR)

Wolf Spider in soil close up
(AR-5-A VWR)

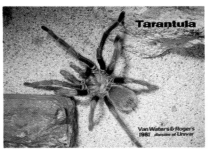

Tarantula adult female
(III-18-W VWR)

Tarantula adult male
(III-18-V VWR)

Jumping Spider
(AR-6-B VWR)

DANGEROUS SPIDERS

Of the nearly 3000 species of spiders that live in the United States, only two groups are considered dangerous to people and both are found in Texas. The better known are BLACK WIDOW SPIDERS; the lesser known are BROWN RECLUSE SPIDERS.

The Black Widow Spider is found outdoors in all kinds of protected cavities. Around homes it prefers garages, gas and electric meters, furniture and many other unbothered places. A Black Widow bite feels like a pin prick and sometimes is not even felt. Usually, a slight local swelling and two red dots surrounded by local redness indicate the location of the bite. Pain becomes intense in 1 to 3 hours and may continue up to 48 hours. Symptoms include abdominal pains, a rise in blood pressure, nausea, profuse perspiration, leg cramps, tremors, loss of muscle tone and vomiting. The toxin also causes breathing difficulties and sometimes unconsciousness.

The Brown Recluse is a nonagressive fellow that spins a white or grayish, nondescript web. Its body and legs cover an area about the size of a quarter, and its color varies from an orange-yellow to dark brown to almost black. The most distinguishing characteristics of this spider are its eyes and its back markings. It has three pairs of eyes arranged in a semicircle on the forepart of the head. The eyes also form the base of a violin-shaped marking on its back. It often lives around human dwellings and is found in bathrooms, bedrooms and closets, as well as under furniture, behind baseboards and door facings, or in corners and crevices. It also likes cluttered garages. Sometimes people are bitten while asleep; others may be bitten by spiders that are in stored clothing. Usually the bite causes a stinging sensation and then intense pain. Within 24 to 36 hours, a systemic reaction may occur, characterized by restlessness, fever, chills, nausea, weakness and joint pain. The bite also produced a small blister surrounded by a large conjested and swollen area. The venom usually kills the affected tissue, which gradually sloughs away and exposes underlying tissue. Healing may take 6 to 8 weeks, leaving scars that might require plastic surgery to repair.

If you are bitten by either of these spiders, seek medical aid immediately.

Brown Recluse Spider = Class Arachnida, Order Araneida, Family Loxoscelidae
Black Widow Spider = Class Arachnida, Order Araneida, Family Oxyopidae

---

CONTROL CLUE

Know these spiders and their habits. Kill them on sight.

Black Widow Spider (Underside view)
(C-USDA-215)

Brown Recluse Spider
(III-17-A VWR)

Spittlebugs

Spittlebugs, also known as Froghoppers, appear in the spring and early summer. Drops of undigested sap mixed with air are excreted by this insect, producing the frothy "spittle" that surrounds its body. This white froth is produced presumably to maintain an artificially high humidity which is required for development of the nymphs. The nymphs suck sap for a living and if infestations are severe, they can cause damage to host trees.

Adults resemble leafhoppers and fly actively during summer. They are yellow to grayish-brown in color, are wedge-shaped, and are about ½ inch long. Of course, the spittle-piles are the key to determining their presence on certain pines, pecans and other trees.

These critters act like they're always mad at somebody, but if I had to live in a pile of spit, I suppose I wouldn't be very happy either.

Spittlebugs = Order Homoptera, Family Cercopidae

CONTROL CLUE

Although these insects are becoming more common in Texas, they are not known to cause significant damage to pecan trees; however, their incidence on pine trees is cause for more concern. If they become a problem, spray with Isotox or Malathion. Read the label.

Pine Spittlebugs on pine
(C-USDA-145)

Spittlebugs on pecan nuts
(C-USDA-167)

Spittlebug adult close up
(HF-1-B VWR)

_____ Squash Bug _____

When "looks" were passed out, this bug was behind the door. In addition to being the favorite contender-to-win in any *ugly* contest, the Squash Bug sucks juices from leaves and stems of plants of the cucurbit family. Squash bugs prefer squash, but will readily feast on pumpkins and melons.

Adults are brownish-gray to dark gray in color and range from ⅝ inch up to one inch in length. Nymphs usually have a green abdomen with crimson head, thorax, legs and antennae when first hatched, but later turn grayish-white with nearly black legs and antennae. You are likely to find a batch of these along with the adults.

Squash Bug feeding causes rapid wilting and leaves soon become blackened, crisp and dead. Some people think these critters are stinkbugs, because when crushed they emit a strong, disagreeable odor. Not so! I mean, they stink, but they aren't stink bugs.

Squash Bug = Order Hemiptera, Family Coreidae

Squash Bug on leaf
(Charles L. Cole)

Stinkbugs

Stinkbugs are one of the most prevalent pests in Texas. Several species attack vegetable, field and fruit crops. They are sucking insects that prey upon beets, okra, squash, beans, peas, corn, tomatoes, and many weeds. Damage is caused by nymphs and adults sucking sap from pods, buds, blossoms and seeds. If the fruit is attacked at an early stage of development, "catfacing" (deformity) or pitted holes will occur on bean pods, tomatoes and squash. These feeding spots are surrounded by hard callouses, causing the fruit to become distorted and dimpled. If your tomatoes turn up with zits and your okra pods grow in a curlicue like a pig's tail, most likely you have a case of stink bugs.

Adults are approximately ½ inch long, triangular shaped and are likely to be either brown or green in color. Adults are winged and they do fly. Crush one and you will know how the stinkbug got it name. Yuk!

Stinkbugs = Order Hemiptera, Family Pentatomidae

CONTROL CLUE

In my vegetable garden, I have the best luck with a Pyrethrins spray. No waiting period after use. Wash your produce and eat it the same day. Read the label. Also, my electric Bug Biter does a super job on them.

Brown Stink Bug
(C-USDA-48)

Southern Green Stink Bug
(C-USDA-47)

Several species of Stink Bugs that attack vegetable gardens
(Ento.-TAEX)

PREDACEOUS STINKBUGS

Stinkbugs have a tarnished reputation as a group, but not every stinkbug is a bad guy. Take a look at Sir Lancelot squaring off with one of the villains: here's a Predaceous Stinkbug attacking a green stinkbug. As the saying goes, one picture is worth a thousand words. Need I say more about the predaceous stinkbug? If you're a bettin' man, let me give you this tip. Put your dough on the little red-and-black guy with the needle nose. I got it straight from the stinkbug's snout.

Predaceous Stinkbugs = Order Hemiptera, Family Pentatomidae

_____ **BENEFICIAL**_____

Predaceous Stinkbug attacking a green stinkbug
(C-USSDA-191)

_____Termites_____

DRYWOOD TERMITES

Drywood and Subterranean Termites are the most destructive insect pests of wood, causing over $1 billion in damages each year in the United States.

In Texas, Drywood Termites have been found in the southern regions from Beaumont to the Rio Grande Valley and inland to the San Antonio-Austin area. Damaging populations seem to be confined to counties along the Gulf Coast, primarily in the Rio Grande Valley northward through Corpus Christi and into the Houston-Galveston-Beaumont areas. Drywood termite damage has been recorded as far inland as San Antonio, Austin and Boerne.

The surest evidence of drywood termite infestation is the presence of fecal pellets. Drywoods spend their entire lives inside wood and seldom is there any visible evidence of their presence. They construct round "kick holes" in infested wood through which the fecal pellets are eliminated from the galleries or tunnels. The pellets accumulate in small piles below the kick holes or will be scattered if the distance between the kick hole and the surface below is too great. These pellets are distinctive and are used for identification. They are hard and elongated, are less than 1/25 inch long, have rounded ends with six flattened or concavely depressed sides with ridges at angles between the six surfaces. Wood damage, though seldom found, definitely indicates infestation.

Drywoods are less dependent on a moisture source than are subterranean termites. Along the Gulf Coast, drywood termites are commonly found in fence posts and pilings. They are also commonly found in wood garage doors and door frames. Garage interiors, particularly in unfinished garages, should be carefully observed since garage doors are commonly left open for long periods and drywood swarmers may enter. A sound coat of paint will deny access to a given surface, but no effective way has been developed to prevent entry through or under wood shingles.

Drywood Termites = Order Isoptera, Family Kalotermitidae

CONTROL CLUE

Fumigation is the most positive method of drywood termite control. First, a structure is completely enveloped in gasproof, heavy plastic sheeting, then a toxic gas is released into the structure. The process is extremely hazardous and occupants must vacate the premises for 1 to 3 days. You can't handle this fumigation by yourself. Call a pro. Select a certified pest control company that offers this service.

Drywood Termite-alate (winged adult)
(Big State-2)

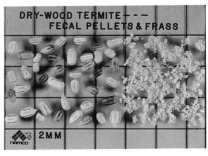

**Drywood Termite fecal pellets and frass —
symptoms of Drywood Termite activity**
(Big State-14)

Drywood Termite fecal pellets
(Big State-25)

**Western Drywood Termite alates and
workers (Prevalent in SE Texas)**
(Big State-23)

**Example of a fumigation treatment
commonly used for large-scale infestations
of the Drywood Termite**
(WP-3)

SUBTERRANEAN TERMITES

Subterranean Termites are found throughout Texas, decreasing in frequency from the Gulf Coast to central regions. In the High Plains and far West Texas, they less frequently attack structures; however, homeowners in all areas of Texas should be on guard.

Subterranean termites derive their nutrition from wood and other material containing cellulose. Paper, cotton, burlap and other plant products often are attacked and consumed. But subterranean termites cannot digest cellulose. They depend on large numbers of one-celled animals (protozoans) that exist in the termite intestine to break down the cellulose to simple compounds that the termites can digest. Termites are highly attracted to odors of wood-decaying fungi.

Moisture is important because subterranean termites dehydrate readily. To survive, they must maintain contact with the soil (their primary moisture source) or other above-ground sources such as structures having defective plumbing or guttering. They must also protect themselves from temperature extremes and from attack by natural enemies such as ants and other insects; consequently, they build shelter tubes. These tubes are constructed by the worker termites from particles of soil or wood and bits of debris held together by salivary secretions.

Dead trees and brush are the original food source of subterranean termites. When land is cleared of this cellulosic material and houses are constructed on these sites, these structures become subject to attack. Termites can enter buildings through wood that is in direct contact with the soil, or by building those shelter tubes over or through foundations, or by entering directly through cracks or joints in and under foundations.

Generally, the first sign of infestation homeowners notice is the presence of swarming reproductives on window sills or near indoor lights. Other indications are the presence of wings, discarded by swarmers as a normal part of their behavior, or those shelter tubes going up the sides of piers, foundations or walls.

Subterranean Termites = Order Isoptera, Families Kalotermitidae, Rhinotermitidae and Termitidae

CONTROL CLUE

This is not a do-it-yourself project. Termite treatment often requires specialized equipment such as drills, pressure injectors, pressure generating pumps, high-gallonage tanks and highly specialized knowledge. Big State Pest Control, based in Houston, also utilizes the services of Tommy, a termite detection dog. This dog is 100% more effective at finding termites than the best trained human inspector. I saw Tommy work in my own home. Superficial treatment is worse than no treatment. Don't compromise; go with a pro.

Subterranean Termite workers, soldiers and reproductives
(VII-3-A VWR)

Earthern shelter tube of subterranean termites
(WP-8)

Termite alate; note dark body of this native subterranean species
(WP-4)

Subterranean Termite workers
(WP-7)

Damage by Formosan Termites to
heartwood of living ash tree
(WP-18)

Alates of Formosan Subterranean Termite
(WP-10)

Carton nest of Formosan Termite used to
store moisture; this extremely hard mass
can be found crusted in wall voids from
floor to ceiling in infested structures.
(WP-15)

Mixed castes of the Formosan Termite,
an introduced species that was first
identified in the United States through the
efforts of Bill Spitz, Big State Pest Control,
Houston, Texas. This termite was
discovered infesting a dock-side
warehouse at the Port of Houston in 1964
and since that time has been found in other
areas of Texas and Louisiana.
(WP-14)

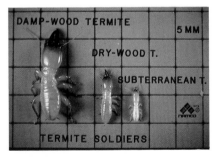

Dry-Wood/Damp-Wood/Subterranean
Termite soldier comparison
(VII-2-L VWR)

_____Thrips_____

Thrips are barely visible insects, less than 1/25 inch long. Many gardeners suffer thrip damage and never determine the cause — you can hardly see these darn things.

These tiny, slender, spindle-shaped, rather active insects vary from pale yellow to yellowish-brown to black in color. Four slender wings are present on the females and these wings are fringed with comparatively long hairs on back margins. Males are wingless. Larvae resemble adults, but have no wings and are smaller. And, try to figure this one out . . . adult females can reproduce regularly without mating with the rarely-found males (somebody really got a lousy deal here).

Thrips rasp the plant tissue and drain the exuding sap. This action causes stunted and deformed plants and when thrip populations are excessive, they can actually kill the plant. Thrips are general feeders, attacking vegetables, flowers and field crops.

Thrips = Order Thysanoptera, Family Thripidae

┌─────────── **CONTROL CLUE** ───────────────────────────┐

In the vegetable garden use Malathion or Diazinon.

└──┘

This is a Thrip
(Ento.-TAEX)

Greenhouse Thrips — adults, fecal material and damage
(TY-2-G VWR)

Greenhouse Thrip adult magnified
(TY-2-J VWR)

Greenhouse Thrips damage to Hypericum groundcover
(TY-2-A VWR)

Thrips as you might see them on a white blossom
(ORTHO-49)

_____Ticks _____

Most Texans, at one time or another, will experience the aggravation of tick bites. Knowledge of ticks and their habitats, as well as methods of tick control, can help you avoid this problem. Ticks are not insects, but are closely related to mites, spiders and scorpions. They are grouped into two families: HARD TICKS — which have a hard, smooth skin and an apparent head; and SOFT TICKS — which have a tough, leathery, pitted skin and no apparent head.

Some hard ticks are more of a problem in Texas. Hard ticks usually mate on a host animal. The female then drops to the ground and deposits from 3,000 to 6,000 eggs which hatch into larvae or "seed ticks." Larvae climb nearby vegetation where they collect in large numbers while waiting for a host to pass within reach. After a blood meal on the host, the engorged larvae drop to the ground, shed their skins (molt), and emerge as nymphs. The nymphs locate a host, engorge themselves with blood, drop to the ground, molt and become adults, and repeat the reproductive cycle.

The term "wood ticks" is applied to several species of hard ticks so similar in appearance and habits that it is difficult to distinguish one from another. In Texas, the most common of these are American Dog Tick; Brown Dog Tick; and Lone Star Tick.

Adult AMERICAN DOG TICKS are chestnut-brown with white spots or streaks on their backs. Unfed adults are about ⅛ inch long; engorged females become slate-gray and may expand to a length of ½ inch. These ticks are widely distributed over the eastern two-thirds of Texas, but are the most abundant in coastal or other humid areas. They are attracted by the scent of animals, so are most often encountered near roads, paths, trails and recreational areas. Although present the year round, American dog ticks are usually most numerous in the spring. Larvae and nymphs feed mostly on small rodents, while adults feed on dogs, cats, *humans* and other animals.

Adult BROWN DOG TICKS are reddish-brown. Unfed adults are 1/8-to-3/16-inch long; engorged females are about ½ inch long. They feed almost exclusively on dogs where they attach to the ears and between the toes. *They rarely attack man or other animals.* Inside the home, the ticks hide behind baseboards, window curtains, bookcases and cabinets, as well as inside upholstered furniture and under the edges of rugs. Outdoors, they hide near building foundations, in crevices between porch flooring, siding, and also beneath porches.

Adult LONE STAR TICKS are various shades of brown or tan. Females have a single silvery-white spot on their backs and males have scattered white spots. Unfed adults are about ⅛ inch long, but after feeding, females may be ½ inch long. Larvae and nymphs feed on wild animals, birds and rodents, while adults feed on larger animals. *All three stages will bite humans.* These ticks live in wooded and brushy areas of Texas, but are most numerous in underbrush along creeks and river bottoms. Lone star ticks are present throughout the year.

Because tick movements and bites are seldom felt, careful and frequent examination for ticks on the body and clothing is imperative. Early removal is important since many disease organisms are not transferred until the tick has fed two to eight hours. Always remove the tick with its mouthparts intact; hasty removal of an attached tick may break off the mouthparts and if they are left in the skin, they can transmit disease organisms or cause secondary infection. To relax tick mouthparts for easy removal, touch the tick with a hot needle or with a drop or two of camphor, alcohol, turpentine, kerosene or chloroform. Sometimes the best method is to grasp the tick firmly with tweezers or fingers and remove it with a slow, steady pull. Always treat the wound with a germicidal agent.

Ticks = Class Arachnida
American Dog Tick, Brown Dog Tick, Lone Star Tick = Class Arachnida, Order Acari, Family
Ixodidae

CONTROL CLUE

Treat the home, yard and pets at the same time. Control light infestations on dogs and cats more than 4 weeks old with 5% Sevin dust; light infestations in buildings usually can be controlled with Dursban or Diazinon spray. Read the label.

American Dog Tick
(C-USDA-208)

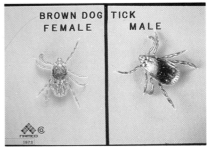

Brown Dog Ticks — male & female
(III-6-C VWR)

____Tomato Pinworm ____

Tomato Pinworm adults are ¼ inch-long gray moths that do most of their business at dusk. Larvae (worms) are light orange at first, but become purplish-black with maturity and will attain a length of ¼ inch. Larval feeding is similar to leafminer damage on tomato plants. Larvae later invade stems and fruit. Invaded tomatoes are useless, even for canning.

So, if you think you have leafminers that seemingly have stopped doodling and have begun housekeeping inside your tomatoes, your problem is likely not loco leafminers, but rather tomato pinworms.

Tomato Pinworm = Order Lepidoptera, Family Gelechiidae

_____ **CONTROL CLUE** _____

The Diazinon you apply to control the other tomato critters should likely handle this one also.

This tomato has been invaded by Tomato Pinworms. Pull it off the bush; even now it is useful only for chunkin' at the neighbor's dog.
(John Norman)

____Treehoppers_____

Treehoppers are small, winged, sucking insects of peculiar and sometimes even bizarre shapes. They live on many plants, but because of their protective color and form, they are usually noticed only when moving. They are closely related to Leafhoppers and although they do suck sap, the primary injury to plants is the result of wounds made by the female; these are double rows of curved slits in the inner bark into which eggs are embedded.

Infested fruit trees, ornamentals and rose bushes look rough, scaly or cracked and seldom make vigorous growth. The fungi that cause rose canker and other diseases gain entrance to the plant through these bark slits.

Treehoppers = Order Homoptera, Family Membracidae

_____CONTROL CLUE _____

Treehopper eggs winter in the wood, so a dormant oil spray will be helpful here. In late spring, eggs hatch into spiny nymphs which drop from the tree and feed on sap of various weeds and grasses until mid-July or August when they become adults, so practice good housekeeping. Clean up! When adults are present on a plant, use an approved Malathion or Diazinon spray. Read the label.

Oak Treehopper female (stem mother)
(HT-3-C VWR)

Oak Treehopper adult close up
(HT-3-J VWR)

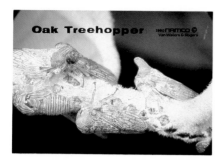

Oak Treehopper nymph, adult male, and adult females
(HT-4-A VWR)

Oak Treehopper adult male, females and damage
(HT-4-B VWR)

Oak Treehopper damage — egg laying scars & feeding punctures
(HT-3-D VWR)

Oak Treehopper damage to a branch
(HT-3-B VWR)

Buffalo Treehopper nymph
(HT-2-D VWR)

Buffalo Treehopper adult
(HT-2-A VWR)

Plant Hopper hiding from his wife on an
azalea bush
(C-USDA-127)

Alfalfa Hopper taking a sun bath
(Ento.-TAEX)

Three-Cornered Alfalfa Hopper — a
country cousin visiting for the summer
(C-USDA-39)

Walking Sticks

The Walking Stick is a defoliator of broadleaf trees. Earlier we mentioned this fellow as sometimes being confused with the praying mantid. These slender, wingless, stick-like insects are pale green when

young, but gradually change to dark green, gray or brown at maturity. The adult female measures up to three inches in length and is more stout-bodied than the male.

Mating usually takes place in August; eggs are dropped to the ground where they overwinter in leaf-litter. In Texas, walking stick eggs normally hatch the summer after they are laid, usually starting in mid-May. The newly hatched walking stick looks like a pint-sized adult.

Walking Sticks = Order Orthroptera, Family Phasmatidae

CONTROL CLUE

At times, walking stick populations build in sufficient numbers to defoliate trees over large areas, but this is the exception, not the rule. Insecticide application is normally not practical. Parasitic wasps and flies are active against immature walking sticks and perform a degree of population control; however, flocks of robins, blackbirds and grackles have a much greater impact. So, walking sticks are strictly for the birds and let's just leave the matter there.

Walking stick from East Texas
(TFS-Pase)

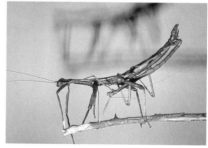

Some Walking stick hanky-panky discovered in Bastrop, Texas
(TFS-Pase)

Wasps

PARASITIC WASPS

There are thousands of species of Parasitic Wasps. Adults are ordinarily minute or even microscopic in size, and that's the reason they pass unnoticed and mostly unappreciated. These darned half-pints are hard to see. Adults usually have black or metallic colors on the body.

Adult wasps lay eggs in the bodies of other insects which might be at various stages of development. After hatching, these wasp larvae feed inside the body of their host. These parasites usually reduce the feeding of or vitality of the host and in due time will kill it. An excellent example of this activity is the tiny Braconid Wasp setting up housekeeping on a hornworm. So, it's a lousy lunch, but it's excellent pest control.

Parasitic Wasps = Order Hymenoptera, Family Braconidae

_____ BENEFICIAL_____

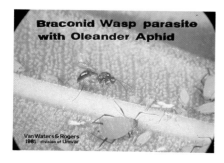

Braconid Wasp parasite with oleander
aphids
(100-1-UU VWR)

Braconid Wasp parasite stinging an
oleander aphid
(100-1-VV VWR)

Braconid parasite cocoons on Tobacco
Hornworm (Adios, hornworm!)
(C-USDA-188)

PREDATORY WASPS

Predatory Wasps are present in a wide variety of colors, shapes and sizes. Solitary and colonial species occur. Adults can be ¼ inch or considerably more in length. Some are stocky in shape; others are svelt and thread-waisted. In the process of stocking food for their larvae, just a few working predatory wasps can remove a significant number of insects from a garden.

These are the wasps most encountered by people. Most species are capable of stinging people and a few aggressively pursue that end if disturbed.

PAPER WASPS build unprotected paper nests out of chewed wood and hang them under eaves or in sheds or in bushes. Most of us call these wasps yellow jackets because they so much favor the true yellow jacket in appearance. True yellow jackets tend to live in much larger groups, usually building their nests underground.

Another wasp pest is the BALDFACED HORNET. This is a large (¾-inch-long) blackish species with white markings. These hornets construct an inverted, pear-shaped, paper carton nest which can be up to 3 feet tall. These nests are usually built in trees, but may be attached to the side of buildings. Baldfaced hornet colonies are populated by hundreds of individuals which are very aggressive when aroused, and their stings can be intensely painful.

Good examples of solitary predatory wasps are the king-sized CICADA KILLER, the MUD DAUBER and the POTTER WASP.

Cicada Killer = Order Hymenoptera, Family Sphecidae
Mud Daubers = Order Hymenoptera, Family Sphecidae
Potter Wasps = Order Hymenoptera, Family Vespidae

CONTROL CLUE

Remember, these are primarily beneficial critters and should be eliminated only if their presence poses a threat to people. CICADA KILLER: Control is rarely warranted. Males cannot sting; females will not sting unless forced to do so. MUD DAUBERS: Nests can be simply removed by hand with a putty knife, as females will not fight back. PAPER WASPS: Wait till dark; wasps will be docile and on the nest. By flashlight, soak the wasps and nest with one of the aerosol stream-sprays containing Pyrethrins. When the wasps are dead, remove the nest. GROUND-NESTING YELLOWJACKETS: Locate the nest during the day, then return after dusk using a red light. Apply about one quart of Sevin spray directly into the nest entrance and then plug the hole with cotton. HORNETS: Nests that pose no threat to humans should be left unharmed; otherwise control and nest removal should be left to a professional.

Yellow Jacket Wasp — adult
(III-13-A VWR)

Cicada Killer Wasp close up
(HC-1-X VWR)

Mud Dauber Wasp adult
(III-15-A VWR)

Bald-faced Hornet — adult
(III-14-A VWR)

Bald-faced Hornet nest in tree
(III-13-C VWR)

Webworm

FALL WEBWORMS

Fall Webworms are frequently confused with tent caterpillars. There are many similarities, but also there are many differences. The loosely woven, dirty-white webs produced by fall webworms are found on the terminal ends of branches. The larvae (worms) consume the foliage within the web. Fall webworm larvae are pale yellow, spotted with black. They attain a length of one inch when fully grown and are covered with long black and white hairs. In most areas of Texas 3 to 3½ generations occur per year.

Pecan trees are the preferred host, but persimmons, hickorys and other trees will also be attacked. Fall webworms have a voracious appetite, so don't ignore them.

Fall Webworm = Order Lepidoptera, Family Arctiidae

CONTROL CLUE

If you regularly grow pecans successfully, you are spraying your trees at appointed times with an insecticide. This regular spraying should control fall webworms along with the other critters. Zolone is an excellent insecticide for this purpose. Webworms can also be burned if you can reach the web, but be careful with the fire.

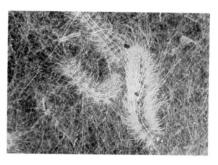

Mature caterpillars of Fall Webworm
(HI-28)

Fall Webworm adult and egg mass
(HI-29)

Fall Webworm larvae

Van Waters & Rogers
1982 division of Univar

Fall Webworm Caterpillars in nest
(LM-7-B)

Fall Webworm Caterpillars and damage
(HI-27)

Fall Webworm nest and damage in tree
(C-USDA-157)

PINE WEBWORM

Pine Webworms derive their common name from the habit the larvae have of weaving clusters of pine needles together in a silken "nest." In Texas, pine webworms feed on loblolly, shortleaf, slash and longleaf pines, and young seedlings frequently die because of being completely defoliated. Infestations on ornamental pines around homes are unsightly.

One to three generations develop per year in East Texas. Adult parent moths appear in late spring and mate, but are seldom noticed because of their small size (about 1 inch wingspan) and their nondescript coloring. When the larvae hatch, groups of up to 75 wander among the needles spinning silken threads. Each larva then bores into a needle and mines it.

Once they have grown too large to feed in individual needles, colonies of larvae feed among loosely webbed clumps of foliage, filling the webbing with oblong, brown fecal pellets. It is an unsightly mess to say the least. This mass of webbing and you-know-what may be 2 to 5 inches long. Full grown larvae (worms) are about ¾ inch long, and are yellowish-brown with dark brown longitudinal stripes on each side of their bodies.

Pine Webworm = Order Lepidoptera, Family Pyralidae

Web and frass of the Pine Webworm
(PD-11)

SOD WEBWORMS

Sod Webworms are the larval (worm) stage of lawn moths. Activity is most apparent during the early evening hours when moths can be seen fluttering above lawns as they drop eggs into the turf. Adults are small and vary in color from white to shades of gray. While at rest, adult moths hold their wings folded close to their bodies, making the snout-like projections on the fronts of their heads even more noticeable. The slender larvae may reach ¾ inch in length and are characterized by a light brown color with several rows of dark spots along the entire body.

During the summer months, sod webworm larvae live on the soil surface in silken tunnels constructed in the thatch of the grass. Damage occurs as larvae chew off grass blades and retreat into these tunnels to consume the foliage. Injury first appears as small brown patches of closely clipped grass. Lawns are particularly susceptible during the months of July and August when temperatures are hot and lawns are not growing vigorously. Large areas may be damaged rapidly if controls are not applied. Sod webworm larvae feed primarily at night and prefer areas in lawns that are hot and dry during daylight hours. Steep slopes, banks and other areas difficult to water properly are most subject to attack. Heavily shaded areas are seldom invaded.

Sod Webworms = Order Hymenoptera, Family Pyralidae

CONTROL CLUE

If three or four sod webworm larvae are found within a 6-square-inch section of dying sod, chemical treatment is recommended. Larvae are most active on cloudy days or at night. Prior to treatment, mow the lawn and rake infested areas to remove dead grass and plant debris. Use Sevin or Diazinon. If a granular form of either insecticide is your choice, a light watering after application will aid in achieving control; if you prefer an emulsifiable concentrate or soluble powder form, water the lawn thoroughly before application.

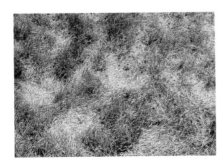

Sod Webworm damage to lawn grass
(ORTHO-46)

Severe Sod Webworm damage to lawn
(Ento.-TAEX)

VEGETABLE GARDEN WEBWORMS

Two brands of Webworms give Texas vegetable gardens fits — Garden Webworms and Beet Webworms. Both are caterpillars and the parents are moths.

GARDEN WEBWORM adults are colored buff with shadings and have irregular light and dark gray markings. Generally they are active at night and are attracted to lights. Larvae (worms) are about an inch long, are yellowish or greenish in color with a light stripe down the back. They sport three dark spots on the side of each segment and these dots form a triangle. Larvae feed primarily on the underside of leaves, skeletonizing them. They spin webs and draw additional leaves into the web as more food is needed. They are general feeders, attacking principally beans and peas. Some may be found during nearly every season in the Lower Rio Grande Valley.

BEET WEBWORM adults are brown moths mottled with lighter and darker spots. Larvae are about two inches long, are slender and are colored yellowish to green with a dorsal black stripe. They also web leaves and devour foliage, often migrate like armyworms and leave behind stripped crops. Beet webworms attack beets, cabbage, beans, peas, carrots, spinach and other crops.

Webworms = Order Lepidoptera, Family Pyralidae

Garden Webworm sitting for portrait
(Ento.-TAEX)

Garden Webworm on vegetable plant
(Ento.-TAEX)

A Diazinon spray or Sevin dust should get 'em. I've not had to use chemicals in my garden to control this pest, because my electric Bug Biter does such a good job on those night-flying adult moths.

_____Weevils_____

Weevils are a big family. They are small, beetle-like critters with mouth parts modified into a downward-curving beak or snout. Weevils attack potatoes, tomatoes, turnips, carrots and many other vegetable, nut, fruit, grain and field crops. Several varieties do dirty work.

VEGETABLE WEEVILS are a variety that go after many Texas garden crops. Both larvae (grubs) and adults feed on the plants, primarily at night. Damage may resemble that of cutworms.

The PEPPER WEEVIL specilizes in bell peppers. Adults are black, about ⅛ inch long with a sparse covering of tan to gray hairs. There may be five to eight generations which appear to be continuous in the Lower Rio Grande Valley. And bell peppers are an important crop down there. Young larvae tunnel into the seed mass in the center of the pepper pods.

CARROT WEEVILS attack not only carrots, but also parsley and dill. Several generations occur in a single season and are a major pest, especially in South Texas. Grubs feed on the exterior or burrow into carrots with damage more likely occurring near the top. After weevils get done with a carrot, even Bugs Bunny wouldn't have it.

Weevils = Order Coleoptera, Family Curculionidae

Insecticides must be applied to kill the adult weevils before they lay eggs. Diazinon, applied thoroughly, will do a good job on brother weevil.

Vegetable Weevil — adult
(C-USDA-56)

Vegetable Weevil — larva
(C-USDA-55)

Vegetable Weevil adult close up
(CW-8-A VWR)

Acorn Weevil close up
(SC-43)

PECAN WEEVIL

Pecan Weevils are about ⅜ inch long and are brown. Adult females lay eggs directly in a pecan after making a hole with that long snout. Grubs that hatch develop rapidly as they feed inside the nut. They are creamy-white with reddish-brown heads. When the nuts drop to the ground in the fall, the developing grubs gnaw ⅛ inch holes through the shell. The grubs crawl through the holes and burrow into the ground.

It takes two or three years before adult weevils emerge again and attack nuts. In other words, this year's weevils will produce the ones you see two years from now. Pecan weevils are late season pests in several areas of Texas. In years when severe infestations occur, this son-of-a-gun may destroy a major portion of the area pecan crop.

Pecan Weevil = Order Coleoptera, Family Curculionidae

CONTROL CLUE

Begin checking trees in the first week in August to determine the presence of weevils. Weevil feeding may cause premature nut drop during the water stage of nut development. Homeowners who regularly spray their pecan trees with insecticide will also control the pecan weevil. Zolone does a good job for me.

Pecan Weevil — adult
(C-USDA-165)

Pecan Weevil — larva
(C-USDA-164)

REPRODUCTION (PINE) WEEVILS

Adults of the PALES WEEVIL and the PITCH-EATING WEEVIL are attracted to stumps and dying trees where they deposit their eggs, but they also feed upon and destroy seedlings in the area. The DEODAR WEEVIL normally breeds in the main stem of dying pines and in the past has been of little concern. In recent years, however, this weevil has been found infesting leaders of apparently healthy saplings, and also it is disposed to attack and destroy seedlings weakened by almost any cause. These weevils are formidable pests in the areas of Texas where pine trees grow.

The pales weevil and pitch-eating weevil look much alike; adults are black or nearly so and are often speckled with whitish markings. They are ¼ to ⅓ inch long. The deodar weevil is slightly less than ¼ inch long and is a rusty-red with two distinct whitish blotches on the back. Females of the three species place their eggs in small holes in the bark of stumps and dying trees. The larvae are small, white, legless and have shiny brown heads. They make separate winding tunnels between the inner bark and sapwood. When fully grown, each larva constructs a conspicuous, oval "chip cocoon," ⅓ inch long, of finely shredded sapwood. Here it pupates and then, as a young adult, chews a round, BB-size exit hole through the cocoon and bark.

Most damage by the pales and pitch-eating weevils occurs in spring and fall. The adults are attracted to the odor of freshly cut pine stumps, scorched pines or dying trees and they feed on the tender bark of nearby seedlings, causing significant damage. The deodar weevil breeds beneath the bark of saplings and larger pines that are dying from competition, suppression or bark beetle attack, but on occasion it attacks young seedlings which may be killed by the larvae tunneling in the cambium and woody tissue, or by multiple bark punctures caused by adult feeding.

Weevils = Order Coleoptera, Family Curculionidae

_____ **CONTROL CLUE** _____

Prompt removal and disposal of pine stumps and of suppressed and dying pine trees is the best control for homeowners. You probably would be inclined to perform this "housekeeping" chore anyway, weevils notwithstanding.

Comparison of adults: Pitch-eating Weevil
(larger); Pales Weevil (middle) Pissodes
(smaller)
(RW-1)

Pales Weevils — adults
(RW-5)

Pales Weevil on pine
(C-USDA-148)

Pales Weevil larva
(RW-3)

Pitch-eating Weevil adults
(RW-6)

Pine regeneration killed by Pissodes
weevils
(RW-9)

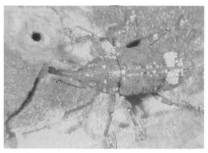

Deodar Weevils (Pissodes), pre-copulation on slash pine billet
(TF-17)

Deodar Weevil (Pissodes) and feeding pit on slash pine
(TF-21)

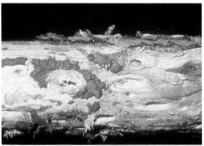

Deodar Weevil (Pissodes) larva
(C-USDA-155)

Weevil chip cocoons beneath bark of pine stem
(RW-10)

SWEET POTATO WEEVIL

This critter, the Sweet Potato Weevil, is worse than an Internal Revenue Agent for being persistent. This ugly devil never quits. Both adults and larvae damage the plants. Adults not only will feed on leaves and vines, but will also feed on, and even breed in, stored sweet potatoes. In the garden, adults lay their eggs on the plant near the soil surface. The grubs that hatch burrow into the vines and tunnel into the roots. There may be as many as eight generations in a single year, and they *don't* hibernate for the winter. They will subsist on weeds, especially wild Morning Glory, through winter.

Adults are about ¼ inch long and are ant-like in appearance. Their head, snout and wings are a dark metallic blue and their thorax and legs are bright orange. Another distinctive characteristic is that very long snout. Larvae are legless grubs almost ½ inch long when full grown. They are white with brown heads.

Sweet Potato Weevil = Order Coleoptera, Family Curculionidae

_____ **CONTROL CLUE** _____

Don't plant sweet potatoes for a couple of years and light a candle. Seriously, there are no chemical controls for this pest. Sorry!

Sweet Potato Weevil
(C-USDA-100)

_____ Whiteflies _____

Whiteflies are very tiny, snow-white insects that resemble moths if viewed under a magnifying glass. If viewed without magnification, they look more like flying dandruff. Adults are about 1/16 inch in length, have four wings and are covered with a white waxy powder. Nymphs (babies) are light green, oval, flattened and are about the size of a pin head. Their bodies are covered with radiating, long filament-like threads. You could confuse them with soft scale insects. Both nymphs and adults feed by sucking plant juices. Heavy feeding can give plants a mottled look, can cause yellowing and even death.

Overlapping generations occur in the Lower Rio Grande Valley during spring, summer and fall. In Texas, whiteflies thrive on gardenias, privet and greenhouse plants, but also will attack tomatoes, potatoes, eggplant, pepper, sweet potatoes and citrus. Sticky honeydew excreted by these critters glazes upper and lower leaf surfaces permitting development of black sooty mold fungus.

Whiteflies = Order Homoptera, Family Aleyrodidae

_____ CONTROL CLUE _____

Malathion or Diazinon applied to control other vegetable pests will ordinarily prevent 'ol whitefly from getting a toe-hold, but if he gets a head start on you, spray twice or more per week. Remember, you will likely see flying "dandruff" first and later the black sooty mold.

Whitefly adults, side, dorsal and ventral views close up
(HW-1-A)

Greenhouse Whitefly
(101-1-M)

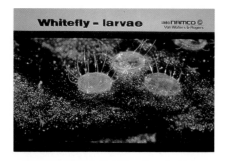

Whitefly larva magnified
(HW-1-E)

Citrus Whitefly adult close up
(HW-3-B)

Citrus Whitefly larva & pupa close up
(HW-3-C)

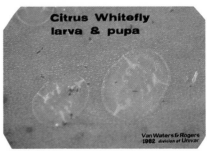

Citrus Whitefly larva & pupa magnified
(HW-3-D)

Black sooty mold on leaves
(100-1-W)

White Grubs

This whole family is trouble. Adults are called May or June Beetles; larvae are called Grub Worms. Adults will chew up leaves; grubs will feed on roots and other underground plant parts of many vegetable and garden crops. As many as 100 species may cause damage to vegetables.

People generally don't like these critters because they are so creepy. They really won't hurt you, but can cause you to hurt yourself. Hold an adult in your hand and it feels like he's eating to the bone. (What you really feel is not teeth chewing, but feet scratching.) Let one sail into your wife's new hairdo (the adults do fly) and at the very least it will mean another trip to the beauty shop ... or maybe to the therapist, mental or physical.

White Grubs = Order Coleoptera, Family Scarabaeidae

CONTROL CLUE

Chickens eat 'em, but who has chickens anymore! Those electric bug killers like my Bug Biter do a super job on the adults; or work some Diazinon granules into your garden soil according to label directions to get the grubs. Like I told you ... trouble!

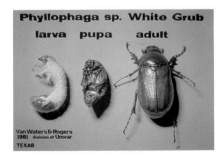

White Grub larva, pupa and adult
(CJ-6-A VWR)

White Grubs in soil close up
(CJ-8-L VWR)

White Grubs in soil and held in hand
(CJ-8-K VWR)

June Beetle (LeConte adult)
(CJ-2-A VWR)

Green June Beetle — larva
(C-USDA-84)

Ten-lined June Beetle adult male
(CJ-1-A VWR)

Green June Beetle on peach
(C-USDA-170)

White Grub turf destruction
(CJ-8-I VWR)

Wireworms

Every gardener who has tilled the soil will have turned up some of these critters from time to time. Wireworms attack virtually every garden crop, damaging planted seed and plant roots. They will also bore into large roots, stems and tubers causing damage ranging from poor stands to complete loss.

Wireworms are larvae of Click Beetles. They come from a large family — over 800 species in North America. Wireworms, cylindrical and elongate in shape, are smooth, shiny and hard-bodied. Color varies from yellow to brownish. When you encounter one of these, kill it.

Note: If you want to know why Click Beetles are called Click Beetles, place one on a hard, smooth surface. Flip him over on his back and wait a minute.

Wireworms (Click Beetles) = Order Coleoptera, Family Elateridae

CONTROL CLUE

Squeezing between thumb and forefinger works nicely. He won't bite. Popping wireworms isn't as much fun as popping cutworms, but do it anyway. Diazinon granules applied according to directions is another more civilized control. Crop rotation (don't plant the same thing in the same place every year) will also offer a measure of control. Wireworms aren't spectacular as critters go, but they can cause severe damage to vegetables and ornamentals. Watch 'em.

Wireworm in soil near sprouting corn seed
(C-USDA-29)

Wireworm or Click Beetle larva
(CO-7-A VWR)

Click Beetle adult species (brown)
(CO-7-B VWR)

Click Beetle adult species (black)
(CO-7-C VWR)

Other Critters

Critters that did not make the squad because of grades or disciplinary reasons:

Silver Spotted Skipper. This caterpillar will fasten together several leaflets for a case in which it lives and feeds.
(C-USDA-105)

Catalpa Sphinx. Fishermen cherish this caterpillar for bait. It's a big 'un, about 3 inches long. Big bait = big fish.
(HI-49)

Tussock Moth larvae dress fancy.
(C-USDA-151)

Linden Looper. This critter doesn't belong in this book; I just like the picture.
(HI-39)

Katydids are grasshopper kinfolks; they mostly just make noise.
(OR-1-F VWR)

It looks like a big powderpuff, but it really is a Forest Tent Caterpillar cocoon.
(LM-4-C VWR)

Pear Slug larva and damage on pear leaf. (Slugs are UGH).
(HYS-2-E VWR)

Wooly Alder Aphid. Under this mop is an Aphid.
(C-USDA-121)

Yellow Jacket Wasp playing "gotcha" with
a caterpillar. The wasp won.
(Sandi Cole)

Peach Twig Borer larva fooling around on
an apricot pit.
(LM-23-A VWR)

Just another Leafhopper.
(Ento.-TAEX)

Velvetbean Caterpillar. This critter is hot
stuff on soybeans.
(C-USDA-45)

I just want you to see a Rottenwood
Caterpillar.
(C-USDA-161)

Sugarcane Beetle. Beetle with a
sweet-tooth.
(C-USDA-38)

Zebra Caterpillar. This guy eats everything.
(C-USDA-140)

Velvet Ants are parasites of wasps and bees.
(C-USDA-210)

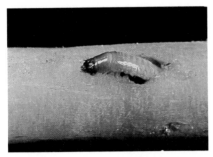

Wireworm. "Here she comes, Miss America . . ."
(C-USDA-52)

Damsel Bugs "make it" off other insects; they are wholly predaceous.
(HE-15-A VWR)

Rice Weevil. You could find this guy in old rice.
(C-USDA-12)

Granary Weevil ruining some wheat.
(C-USDA-21)

These little beetles are often called "bran bugs" or, mistakenly, weevils.
(C-USDA-11)

Cadelle Beetles are bad news in stored grain.
(C-USDA-23)

A Cadelle Beetle kid.
(C-USDA-22)

Look closely, the gray mass is a pack of Cabbage Aphids.
(Ento.-TAEX)

Whitefly adult emerging from pupal case.
(HW-1-H VWR)

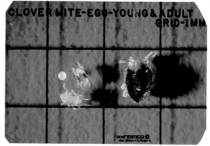

Clover Mite egg, young and adult.
(IV-32-A VWR)

Webworm doing its thing in sorghum.
(C-USDA-86)

Grubs are worse than little kids; here's one in a sweet potato.
(C-USDA-101)

I have no idea what this is. Do you?
(SC-44)

Appendix

CRITTER GROUPS

The following list is not intended to be an index of critters appearing in this book. True, most are shown in living color and are discussed in some detail, but as in the cast of a play, their frequency of appearance and importance to the plot will vary. Aphids are a good example. Aphids are like kittens ... they can do mischief in many situations; borers, varieties notwithstanding, generally follow the same script if attacking a pecan tree or an oak tree. And critters are not necessarily villains; many are heroes and their presence should be encouraged. So, when you have a problem, perhaps this critter-cast will be a clue as to where to look for the culprit.

BENEFICIAL CRITTERS

Preying Mantids

Assassin Bugs

Minute Pirate Bugs

Ladybird Beetles

Ground Beetles

A Soft-Winged Flower Beetle

Syrphid Flies

Lacewings

Ant Lion (Doodle Bug)

Honeybee

Wild Bees

Parasitic Wasps

Predatory Wasps

Spiders

Bark Louse

Mealybug Destroyer

Predaceous Stink Bug

Tiger Beetle

Sucking
- Aphids
- Leafhoppers
- Scale Insects
- Whiteflies
- Spider Mites

Chewing
- Thrips
- Leaf & Flower Feeding Beetles:
  - Rose Chafers
  - Rose Leaf Beetles
  - Twelve-Spotted Cucumber Beetle
- Rose Slugs
- Leafrollers
- Leaf-Cutter Bees
- Rose Stem Borers
- Gall Wasps
- Rose Midge
- Grasshoppers
- Leaf & Flower-Feeding Larvae

_____COMMON HOUSE-INFESTING ANTS_____

Species	Nest Location	Preferred Food
Argentine	Lawns & plant beds	Sweets, animal fat
Crazy	Trash piles, rotten wood, soil, cavities in trees	Sweets, meat, grease, fruit
Fire	Lawns, plant beds, gardens	Meat, grease
Little Black	Lawns, under objects, rotten wood	Grease, sweets, meat, fruit
Pavement	Cracks in paving	Grease, meat, honey
Pyramid	Gardens & plant beds	Sweets
Acrobat	Wood & mortar	Meats & sweets
Carpenter	Stumps, logs, fences, homes	Sweets and nearly everything else
Thief	Nests of other ants, soil, cracks in walls	Meat, sweets, cheese-grease
Odorous	Under floors, stones, or in walls	Sweets, meat, dairy products
Pharaoh	Various locations near heat and moisture sources	Meats, grease, sweets

Whole Grain Insects
Angoumois Grain Moths
Grain Weevils:
  Rice Weevils
  Granary Weevils
Bean & Pea Weevils
Cadelle Beetles

Miscellaneous Pests
Brown Spider Beetle
Mealworms
Psocids (Booklice)

Processed and Dried-Food Insects
Grain & Flour Beetles,
Cigarette & Drugstore Beetles:
  Confused Flour Beetles
  Saw-Toothed Grain Beetles
  Cigarette Beetles
Dermested Beetles:
  Larder Beetles
  Carpet Beetles
  Cabinet Beetles
Flour Moths:
  Indian-Meal Moth
  Almond Moths
  Mediterranean Flour Moth

Attacking the nuts
Pecan Nut Casebearer
Hickory Shuckworm
Pecan Weevil
Stinkbugs & Plantbugs

Attacking the limbs, trunk and twigs
Pecan Twig Girdler
Red-Shouldered
  Shot-Hole Borer
Flatheaded Borers
Obscure Scale

Attacking the foliage
Aphids
Mites
Pecan Leaf Casebearer
Pecan Nut Casebearer
Pecan Cigar Casebearer
Pecan Phylloxera
Sawflies
May Beetles
Fall Webworm
Walnut Caterpillar
Pecan Catocala
Pecan Spittlebug
Pecan Bud Moth
Leafminers

Conifers
- Southern Pine Beetle
- IPS Engraver Beetle
- Black Turpentine Beetle
- Nantucket Pine Tip Moth
- Reproduction Beetles
- Texas Leaf-Cutting Ant
- Pine Sawflies:
  - Red-Headed pine sawfly
  - Black-Headed pine sawfly
- Pine Webworm
- Bagworms
- Pine Colaspis Beetle
- Sawyer Beetles
- Ambrosia Beetles
- Termites
- Scales

Hardwoods
- Fall Webworms
- Tent Caterpillars
- Variable Oak Leaf Caterpillars
- Elm Leaf Beetles
- Walnut Caterpillars
- Cankerworms
- Twig Girdlers
- Leaf Rollers
- Carpenter Worms
- Carpenter Ants
- Carpenter Bees
- Walking Stick
- Aphids
- Borers
- Termites
- Gall Insects
- Scales
- June Beetles
- Catalpa Worm
- Powder Post Beetles

Bark Beetles:
- Southern Pine Beetles
- IPS Engraver Beetles
- Black Turpentine Beetles

Wood Borers:
- Southern Pine Sawyers
- Turpentine Borers
- Ambrosia Beetles

Root and Twig Insects

Bark Feeders:
- Nantucket Pine Tip Moth
- Pine Weevils
- White Grubs

Leaf Feeders:
- Red-Headed Pine Sawfly
- Black-Headed Pine Sawfly
- Sawfly
- Pine Webworms
- Texas Leaf-Cutting Ant
- Pine Colaspis Beetle

Miscellaneous Insects:
- Pine Needle Miner
- Pine Pitch Midge
- Scale Insects
- Aphids
- Spider Mites

Soil
- Cutworms
- Wireworms
- White Grubs
- Maggots
- Rootworms
- Sweet Potato Weevil
- Mole Crickets
- Pillbugs & Sowbugs

Sucking
- Squash Bug
- Harlequin Bug
- Stink Bugs
- Leaffooted Bugs
- Garden Fleahopper
- Aphids
- Leafhoppers
- Sharpshooters
- Thrips
- Mites
- Whitefly

Chewing
- Corn Earworm
- Cabbage Looper
- Imported Cabbageworm
- Diamondback Moth
- Armyworms
    - Beet Armyworm
    - Fall Armyworm
    - Yellowstriped Armyworm
- Tomato & Tobacco Hornworm
- Tomato Pinworm
- Serpentine Leafminer
- Squash Vine Borer
- Pickleworm
- Melonworm
- Saltmarsh Caterpillar
- Garden Webworm
- Beet Webworm
- Cowpea Curculio
- Vegetable Weevil
- Colorado Potato Beetle
- Golden Tortoise Beetle
- Flea Beetles
- Mexican Bean Beetle
- Cucumber Beetle
- Squash Beetle
- Blister Beetles
- Grasshoppers
- Texas Leafcutting Ants

_____ WOOD-DESTROYING CRITTERS _____

- Powderpost Beetles
- Death Watch Beetles
- Bark Beetles
- Flatheaded Borers
- Old House Borers
- Flat Oak Borer
- Other Roundheaded Borers

- Drywood Termites
- Subterranean Termites
- Timber Worms
- Carpenter Ants
- Carpenter Bees
- Spider Beetles
- Weevils, Snout Beetles

_____NON-CHEMICAL CONTROLS_____

Plant pests are controlled in a number of ways. Mother Nature provides factors that influence insect numbers without any effort by man; then we have biological control ... natural control with an assist by man; then there is what might be called legal control, exclusion by quarantine, to prevent the introduction of critters; also there is control by cultural practices, by mechanical and physical practices; and, finally, by the use of chemicals.

_____ NATURAL CONTROLS _____

Not many insects live in all climates — arctic, temperate and tropical. Cold winter temperatures restrict the range of some insects; summer heat limits others. Some insects prefer a warm, moist climate; others like it warm but dry. Some insects can fly, or can be carried by the wind over long distances; others can crawl only short distances. Geographic barriers, like lakes, rivers and mountain ranges, can check the spread of insects; the nature of the soil deters others, for example, wireworms flourish in poorly drained soil — nematodes in sandy soil. Birds, moles, skunks, snakes, lizards, newts, salamanders and toads eat many insects. Birds consume enough to more than compensate for the strawberries, tomatoes, grapes or blueberries they also eat. Bird netting over your vulnerable fruiting plants will usually tip the scale in your favor. Skunks and armadillos eat a lot of grub worms, but in the process root holes in turf and garden. Skunk and armadillo critters are nocturnal. They can be easily frightened and will leave of their own volition if given the opportunity, but be wary of a skunk if he raises his tail; the hunted then becomes the hunter. Most of us know enough not to mess with a skunk, but your dog might have to learn the hard way. (Note: Tomato juice will neutralize skunk odor, but don't try to feed it to your dog if confrontation results in his getting skunk-sprayed; he won't drink it. Bathe him in it.)

_____ BIOLOGICAL CONTROLS _____

consist of artificial restoration of the balance of nature; Bacillus thuringiensis as a control for certain caterpillars is a good example. Bringing in insect predators and parasites, encouraging birds, introducing certain nematodes or fungi to prey on the undesirables will be effective to varying degrees.

_____ CONTROL BY EXCLUSION _____

can be either legal or voluntary. The Plant Quarantine Act of 1912 provides that whenever it is deemed necessary, in order to prevent the introduction of any dangerous insect or plant disease, the federal government will have the power, after a public hearing, to prohibit the importation, or shipment interstate, of any class of plants or plant products from any country or locality and from any state or territory in this country. Such specific prohibitions are called quarantines. In addition to federal quarantines there are also state quarantines. It was never expected that a quarantine could keep out a certain pest forever, but the expense of the inspection service is justified if the insect is excluded for a period of time sufficient for us to learn its life history, to develop controls and to introduce its predators or parasites.

_____ CULTURAL CONTROLS _____

may simply be a matter of good housekeeping in the garden. Sometimes it does not even require labor, but rather a little planning based on the knowledge of life histories of certain insects. Most insects attack only a small number of related plants. Cabbage worms chew members of the cabbage family; the squash vine borer prefers the cucurbit group; Mexican bean beetles don't like much except beans; sweet potato weevils have an ongoing love affair with sweet potatoes. By switching locations of crops (crop rotation) you can starve out certain pests, or at least keep them from building up huge populations. Commercial farmers have taught us trap-cropping (planting an early, expendable crop to attract an insect pest for elimination). Soil cultivation destroys some critters; tilling the vegetable patch in the winter exposes some larvae and pupae and they are killed by the cold; tilling in the summer kills some by exposure to hot temperatures. Timing of planting is another good cultural control for insects. Green beans planted early often mature between broods of the Mexican bean beetle; early summer squash may come along ahead of the squash vine borer; late corn is less apt to be injured by the European corn borer. Often the character of the foliage keeps off pests. Leafhoppers usually prefer varieties with smooth leaves, but the azaela whitefly restricts operations to varieties that have hairy leaves. The effectiveness of these control-tricks will depend upon your knowledge of and experience with your individual garden . . . further proof that good-gardening is really an art and not a science.

Good housekeeping in the garden is tremendously important in controlling pests. Clean up all plant parts after harvest, compost what you can and burn anything capable of causing trouble later, for instance, the limbs severed by the twig girdler. Get rid of weeds and plant debris to avoid offering a haven to cucumber, Mexican bean and Colorado potato beetles as well as a multitude of other pests. I like to shred all plant matter larger than pine needles before composting.

CONTROL BY MECHANICAL AND PHYSICAL MEASURES

consists of barriers being erected between the plant and the pest. A wire fence will keep out rabbits and other animals; bulbs planted in wire baskets will protect against mice and moles; wire shields will keep dogs off shrubbery; hardware cloth protects orchard trees from rats, mice and squirrels; nylon bird netting properly placed will confound the brightest Texas mockingbird and spare your ripened strawberries, tomatoes and fruits; a cardboard cylinder or tin can will shield young transplants from cutworms. Dare to be ingenious. Remember, all is fair in love, war and gardening.

CHEMICAL CONTROLS
USING PESTICIDES

We have more insecticides, fungicides, herbicides and technology available to us than ever before in our history. We also have the greatest number of safety regulations controlling their use. And we also enjoy the most abundant and varied food supply of any known civilization to date. So, maybe everything within the chemical pest control industry is not all *right*, but certainly everything can't be all *wrong*.

The Federal Environmental Pesticide Control Act of 1972 in part prohibits the application of any pesticide in a manner inconsistent with its labeling. This means that a pesticide cannot be recommended unless it is registered for the specific pest. Such registration is accorded only after intensive scientific investigation is responsibly performed and documented, attesting to the relative safety of said pesticide. In other words, it's a case of being guilty until proven innocent. So, READ YOUR LABEL; don't arbitrarily increase suggested dosages; if your particular pest problem is not described on the label, ask a responsible authority. Don't guess.

Pesticide label clearances are subject to change. Remember, the pesticide user, YOU, are always responsible for the effects of pesticides on your own plants or household goods as well as for problems caused by drift from your property to other property or plants. Always read the container label and carefully follow instructions.

PETROLEUM OIL SPRAYS
(DORMANT, SUMMER AND SUPERIOR OILS)

Petroleum oils are insecticidal in that they suffocate the egg or the immature form or the adult form of pests. They are not toxic to the nervous system of animals, therefore, they are safe for use by man. They are contact insecticides that interfere physically, rather than chemically, with respiration. If the user will read the label and follow directions, few problems will be incurred.

DORMANT OILS are the heaviest and are formulated for use on plants that are dormant. These oils should be applied as late in the dormant season as possible, but before plants enter the bud-break stage in the spring. DO NOT USE A DORMANT OIL DURING THE GROWING SEASON, unless the label expressly states that it can be used.

SUMMER OILS are lighter than dormant oils and are formulated for use during spring and summer. Follow label directions closely. Don't cheat or you will be sorry.

SUPERIOR OILS are the most stringently refined of all and are excellent for horticultural use. They are so named because they are refined under certain specifications which allows them to be used in any season provided weather conditions permit their use. Superior oils are refined primarily for use on trees during the growing season, but they may also be used as a dormant oil.

Certain plants are highly sensitive to oils and should not be treated; these include hickory, certain conifers, ferns, palms and African violets. Many cacti and succulents are also susceptible to injury. Read your label. Most labels recommend that oils be used at temperatures between 40 degrees and 90 degrees F. This is the preferred temperature range. Petroleum oils are probably the most effective of all the pesticides against scale insects. When using oils, spray thoroughly to get coverage of the entire plant; if this is achieved no further spraying should be necessary for several months or a year.

Index

A

Aedes aegypti (tiger mosquito), 124
Aedes sollicitans, 125
Aedes taeniorhynchus, 125
Aedes triseriatus (tree hole mosquito), 124
Aedes vexans, 124
agricultural ants, 8-9
American dog tick *(Arachnida acari ixodidae),* 188-189
angoumois grain moth, 132-133
Anopheles crucians, 125
Anopheles quadrimaculatus, 125
Anoplura pediculidae (head lice), 106-108
ant cows, 13-14
antlions *(Neuroptera myrmelcontidae),* 11-12
ants *(Hymenoptera formicidae)*
   agricultural, 8-9
   carpenter, 1
   cut, 10
   fire, 2-4
   fungus, 10
   harvester, 8-9
   household, 5-6, 223
   night, 10
   parasol, 10
   pharaoh, 7-8
   red harvester, 8-9
   sugar, 7-8
   Texas leafcutting, 10
   town, 10
   velvet, 219
aphids *(Homoptera aphididae),* 13-14
   cabbage, 220
   gall-making, 100-102
   rose-attacking, 150-152
   woolly alder, 217
Arachnida (tick), 187-189
Arachnida acari ixodidae (American dog tick), 188-189
Arachnida acari tetranychidae (spider mite), 120-123
   two-spotted, 151-152
Arachnida acari trombiculidae (chiggers), 69-70

Arachnida araneida loxoscelidae (brown recluse spider), 172-175
Arachnida araneida oxyopidae (black widow spider), 172-175
Arachnida scorpionida (scorpion), 164
Araneida lycosidae (jumping spider), 172-175
Araneida salticidae (wolf spider), 172-175
Araneida theraphosidae (tarantula), 172-175
armored scale insects *(Homoptera diaspididae),* 156-163
army cutworm, 84
armyworm *(Lepidoptera Noctuidae),* 15
asp, 59-60
assassin bug *(Hemiptera reduviidae),* 16

B

bagworm *(Lepidoptera psychidae),* 17-18
baldfaced hornet, 196
banks grass mite, 120-123
bark lice *(Psocoptera pseudocaecilidae),* 19
barnacle scale insect, 157
bees, 20-22
   bumble bee, 21
   carpenter, 21
   honey, 20
   leafcutting
      rose-attacking, 151-152
   wild, 21-22
beet armyworm, 15
beetles
   blister, 23
   borers
      cottonwood, 43-44
      old house, 40-42
      peach tree, 44-45, 218
      rose stem, 151-152
      shothole, 46-47
      squash vine, 47-48
      tree, 48-50
   cadelle, 220
   carpet
      as fabric pest, 89-91
      in pantry, 132-133

drugstore beetle *(Coleoptera anobiidae)*,
132-133
drywood termite *(Isoptera kalotermitidae)*,
180-182

E

earthworm *(Phylum annelida)*, 86-87
earwig *(Dermaptera labiduridae)*, 87-88
earworm, corn *(Lepidoptera noctuidae)*, 77-78
rose-attacking, 152
elm leaf beetle *(Coleoptera chrysomelidae)*,
26-27
eriophyid mite, 120-123
euonymus scale insect, 157-159

F

fabric pests, 89-91
face fly *(Diptera muscidae)*, 95
fall armyworm, 15
fall webworm *(Lepidoptera arctiidae)*,
198-200
field cricket, 79
fire ant, 2-4
firebrat *(Thysanura lepismatidae)*, 167-168
flat grain beetle, 220
flatheaded tree borer *(Coleoptera
buprestidae)*, 48-50
flea beetle *(Coleoptera chrysomelidae)*, 28
fleahopper *(Hemiptera miridae)*, 93-94
fleas, 92-93
flesh fly *(Diptera sarcophagidae)*, 95
flies, 94-100
blow, 95
bluebottle, 95-98
deer, 95-97
fruit, 95-98
gall-making, 100-102
greenbottle, 95
horsefly, 95-97
housefly, 94-97
mediterranean fruit, 95-98
Mexican fruit, 95-98
syrphid, 98-99
tachinid, 99-100
vinegar, 95-97
whitefly, 210-212, 220
Florida red scale, 163
flour beetle *(Coleoptera tenebrionidae)*, 132,
134
flour moth, 132, 135
flower thrip, 151-152
forest tent caterpillar, 217
formosan termite, 185
froghopper *(Homoptera cercopidae)*, 176-177
fruit fly *(Diptera tephritidae)*, 95-98

fruit tree bark beetle, 46-47
fruitworm, tomato, 77-78
fungus ant, 10
furniture carpet beetle, 89-91

G

gall wasp *(Hymenoptera cynipidae)*, 151-152
galls, insect-induced, 100-102
German cockroach *(Orthroptera blattellidae)*,
73-75
girdler, twig *(Coleoptera cerambycidae)*,
38-39
golden tortoise beetle *(Coleoptera
chrysomelidae)*, 29
grain beetle, 132
granary weevil, 219
granulated cutworm, 84
grasshopper *(Orthroptera)*, 103-104
greenbottle fly *(Diptera calliphoridae)*, 95
green lacewing *(Neuroptera chrysopidae)*,
112-113
ground beetle *(Coleoptera carabidae)*, 29-30
grub, 221
white, 213-214
grub worm *(Coleoptera scarabaeidae)*,
213-214

H

hard tick, 187-188
hardwoods, critters in, 225
harlequin bug *(Hemiptera pentatomide)*, 105
harvester ant, 8-9
head lice *(Anoplura pediculidae)*, 106-108
headworm, sorghum, 77-78
Hemiptera anthocoridae (minute pirate bug),
120
Hemiptera coreidae (squash bug), 177-178
Hemiptera lygaeidae (chinch bug), 70-71
Hemiptera miridae
fleahopper, 93-94
plantbug, 66-67
Hemiptera pentatomidae
harlequin bug, 105
stinkbug, 178-180
catfacing, 66-67
Hemiptera phylloxeridae (pecan phylloxera),
140-141
Hemiptera reduviidae (assassin bug), 16
Hemiptera rhopalidae (boxelder bug), 51-52
Hemiptera tingidae (lace bug), 111-112
hickory shuckworm, 108-109
Homoptera aleyrodidae (whitefly), 150-152,
210-212
Homoptera aphididae (aphid), 13-14, 150-152
Homoptera cercopidae (spittle bug), 176-177
Homoptera cicadellide

squash bug *(Hemiptera coreidae)*, 177-178
squash vine borer *(Lepidoptera sesiidae)*,
47-48
stinkbug *(Hemiptera pentatomidae)*, 178-180
catfacing, 66-67
stone cricket, 80
subterranean cutworm, 84
subterranean termite *(Isoptera
kaloterminidae, rhinotermitidae and
termitidae)*, 183-185
sugar ant, 7-8
sugarbeet maggot *(Diptera otitidae)*, 149
sugarcane beetle, 218
sweet potato weevil *(Coleoptera
curculionidae)*, 209-210
syrphid fly *(Diptera syrphidae)*, 98-99

T

tachinid fly *(Diptera tachnidae)*, 99-100
tarantula *(Araneida theraphosidae)*, 172-175
tea scale insect, 162
10 (ten) moth larva *(Lepidoptea saturniidae)*,
59-60
tent caterpillar *(Pepidoptera lasiocampidae)*,
61-62
termite alate, 184
termites, 180-185
drywood, 180-182
subterranean, 183-185
Texas citrus mite, 120-123
Texas leafcutting ant, 10
three-cornered alfalfa hopper, 193
thrip *(Thysanoptera thripidae)*, 186-187
rose-attacking, 151-152
Thysanoptera thripidae (thrip), 186-187
rose-attacking, 151-152
Thysanura lepismatidae (silverfish and
firebrat), 167-168
tick *(Arachnida)*, 187-189
tiger mosquito *(Aedes aegypti)*, 124
tobacco thrip, 151-152
tomato fruitworm, 77-78
tomato pinworm *(Lepidoptera gelechiidae)*,
190
tomato rust mite, 122
tortoise beetle *(Coleoptera chrysomelidae)*, 29
town ant, 10
tree borer, 48-50
tree critters, 225
tree hole mosquito *(Aedes triseriatus)*, 124
treehopper *(Homoptera membracidae)*,
191-193
tuliptree scale insect, 161
tussock moth, 217
twig girdler *(Coleoptera cerambycidae)*, 38-39
two-spotted spider mite *(Arachnida acari
tetranychidae)*, 151-152

V

variable oak leaf caterpillar *(Lepidoptera
notodontidae)*, 63-64
varied carpet beetle, 89-91
vegetable critters, 226
vegetable garden webworm *(Lepidoptera
pyralidae)*, 203-204
vegetable weevil, 204-205
velvet ant, 219
velvetbean caterpillar, 218
vinegar fly, 95-97

W

walking stick *(Orthroptera phasmatidae)*,
193-194
walnut caterpillar *(Lepidoptera
notodontidae)*, 64-65
wasps, 195-198
gall-making, 100-102
rose-attacking, 151-152
paper, 196
parasitic, 195-196
predatory, 196-198
sawfly, 154-155
yellow jacket, 218
wax scale insect, 162
webworm, 198-204, 221
fall, 198-200
weevil *(Coleoptera curculionidae)*, 204-210
carrot, 204
granary, 219
pecan, 206
pepper, 204
reproduction (pine), 207-209
rice, 219
sweet potato, 209-210
vegetable, 204-205
whitefly *(Homoptera aleyrodidae)*, 210-212,
220
rose-attacking, 150-152
white grub *(Coleoptera scarabaeidae)*,
213-214
wild bee, 21-22
wireworm *(Coleoptera elateridae)*, 215-216,
219
wisteria, 161
wolf spider *(Araneida salticidae)*, 172-175
wood-destroying beetle, 40-42
wood-destroying critters, 226
woodpecker, 49
wood tick, 188
woolly alder aphid, 217
worms
earthworm, 86-87
grub, 213-214
webworm, 198-204, 221
wireworm, 219

Y

Z

NOTES

I know, "Don't write in the book," we have been taught;
but do it . . . and do it here!

NOTES